历史·文化·传承

历史·文化·传承

HISTORY CULTURE INHERITANCE

Masterpiece of DHGP Cultural&educational Architectural Design

华谏国际文化教育建筑设计作品专辑

罗凯 著

中国建筑工业出版社

华谏国际

DHGP华谏国际是由德国DHGP建筑设计集团与中国建筑师共同创立的一家全方位多元化的国际性设计公司，在德国柏林和中国上海分别注册。公司从事广泛的设计服务项目，包括全面的建筑工程设计、规划设计、景观设计、室内设计等。

DHGP华谏国际为政府部门、高等院校、知名房地产商完成了众多成功的作品，其作品遍及中国和德国，表现出十分国际化的特色。上海华谏规划建筑设计研究中心和上海华东建设发展设计有限公司拥有境内建筑工程设计资质，为DHGP华谏国际在国内完成优秀设计作品提供了可靠的保障。

DHGP华谏国际成功主要的因素是其建筑设计作品极具历史感和文化性的的感染力，拨动人心弦的精品设计理念和不羁窠臼的创新意识。DHGP华谏国际一向力求达到尽善尽美，并致力于研讨文化教育建筑的表达形式及其内涵。

DHGP华谏国际有能力承担各种规模或范围的设计项目，提供各种规划、建筑设计和施工阶段的服务，并保证有按时完工及控制预算的优良记录。DHGP华谏国际的作品不仅以独特的历史主义风格和创意来体现建筑特色，并且以训练严格的设计团队、精益求精的工作方法来确保独创的设计理念得以实现。

DHGP华谏国际非常关注中国市场，一直积极参与中国的工程项目、DHGP华谏国际在2001—2006年五年时间里，创造了一大批规划及建筑设计的精品。主要大型工程项目有：

华东政法学院松江校区
上海工程技术大学松江校区
复旦大学新江湾城校区
上海理工大学军工路校区
上海电力学院平凉路新校区
上海水产大学临港新校区
上海应用技术学院漕宝路校区
华东师范大学闵行新校区（景观设计）
上海大学东部校区
上海应用技术学院奉贤新校区
上海市宝山气象中心
上海杨浦国际商贸区规划
南京钢铁集团总部办公楼，生产指挥中心（智慧之眼）
上海佘山紫都·晶园别墅区
……

DHGP华谏国际已成为上海文化教育建筑设计的一面旗帜，其独特的规划和建筑设计理念已成就了一个上海文化教育建筑的“DHGP华谏时代”。DHGP华谏国际一直将打造“中国最具历史感和文化性的设计公司”的精品理念作为其设计的核心价值观；“华谏国际历史主义”以其饱含文化和历史感的设计作品和动人心扉的设计理念，在上海乃至中国文化教育建筑设计史上写下了浓重的一笔。

DHGP

DHGP is a multi-oriented international architectural design firm founded by DHGP Germany and Chinese architects. The company is registered in both Dusseldorf and Shanghai, offering a wide range of design services including architectural design, urban planning, landscape design and interior design. DHGP has completed many successful projects for government, universities and prominent real estate developers throughout Germany and China. Their works feature internationalization. DHGP Shanghai and Shanghai East China Construction Development and Design Ltd have obtained the A license for architectural and engineering design from Ministry of Construction P.R.C. which provides reliability for the high-quality design and service in China.

DHGP's success is due largely to its historical and cultural inspirited works. Moreover, their projects exert impressive design fused with unique innovation and advanced concept. DHGP tries to seek the essence of architecture through devoting energy to exploring the form and connotation of educational architecture as well as high-class cultural real estate projects. DHGP has the capability of taking on a wide range of large-scale projects, providing services in planning, architectural design and engineering design in different stages of constructions, with an additional time constraint and budget control guarantee. DHGP is not only defined by its unique historical style and creativity, but also its extensively trained design teams that strive for nothing short of accuracy and originality in each project.

DHGP is focusing its vision on the Chinese market. They are constantly participating in Chinese construction projects. DHGP has created numerous planning and architectural design works in the past 5 years. The main large-scale construction projects include:

Songjiang Campus, East China University of Politics and Law;
Songjiang Campus, Shanghai University of Engineering Science;
New Jiangwan Campus, Fudan University;
Jungong Road Campus, University of Shanghai for Science and Technology;
Pingliang Road Campus, Shanghai University of Electric Power;
Lingang Campus, Shanghai Fisheries University;
Caobao Road Campus, Shanghai Institute of Technology Campus;
New Min Hang Campus (Landscape Design),East China Normal University;
East Campus, Shanghai University;
New Fengxian Campus, Shanghai Institute of Technology Campus;
Shanghai Baoshang Weather Centre;
Shanghai Jiading Juyuan CBD planning;
Shanghai Yangpu International Business Trading Area planning;
Office Building, Nanjing Iron & Steel United Co., Ltd;
Command Centre (Eye of Knowledge), Nanjing Iron & Steel United Co., Ltd;
Shanghai Crystal Palace.

DHGP has become a flagship of Shanghai cultural education in architectural design. Its unique planning and architectural design philosophy have been considered as the 'DHGP Era', most notably recognized in Shanghai cultural education of architecture. DHGP's design philosophy of being 'the most historically and culturally inspired design company' will continue to be the core value of the company. 'DHGP historical style' and its culturally inspired works, together with its design philosophy that appeals to us will become the major architectural trend in the Chinese cultural education of architecture.

Larkin
Chief Architect of ECCSH
Chief Representative of DHGP Germany
General Manager of DHGP Shanghai

罗凯
上海华东建设发展设计有限公司总建筑师
德国华谏国际(DHGP)建筑设计集团中国区首席代表
上海华谏规划建筑设计研究中心总经理

A building will massively stand on a land for a hundred years or more. An architect should learn to take his responsibility. In my mind, conversation and exchange are continuously happening between space and time, inheritation and creation, materials and values. When I once again knock at the door of art, I could not help wandering between emotions and senses, letting the nature brings others and me to its arms. But a large number of buildings are built in the wrong places, lacking of spirituality, charm or temperature. While DHGP is devoted to seeking the nature of life and creating an architectural vocabulary that displays volition, appetency, history and culture.

一幢建筑，她要在坚实的大地上矗立百年甚至更长；建筑师必须要学会善待自己的设计权利。在我眼里，空间与时间、继承和创造、材料和价值，无时无刻不在对话与交流；我游弋在感性与理性之间，一次次叩响艺术之门，任自然裹携着艺术走进我们的生活。DHGP 华谏国际需要创造的正是能表现意志力、亲和力、历史感、文化感的建筑语汇以及对生活的热爱。

目录 Contents

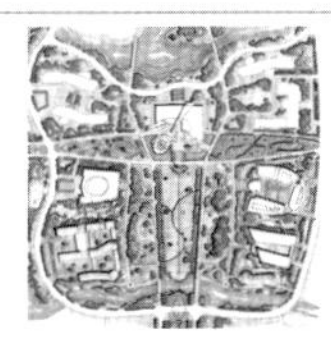

序

罗凯先生带领德国华谏国际建筑设计集团在2001年至2006年间，规划设计了大量的上海市高校文教建设项目，即将由中国建筑工业出版社出版《历史•文化•传承——华谏国际文化教育建筑设计作品专辑》。他希望我为这本丰盛的集子作序，作为我指导过的硕士生，他的盛情难却，我恭敬不如从命。

阅毕全书，掩卷深思，我颇有感触：其一，其以一己之力在短短数年间，承接项目如此之多，在整个上海的高校建筑市场上，占据半壁江山，创造属于他们自己的辉煌，我深感其不易；其二，其从总体规划到单体建筑，从室内设计到室外环境，罗凯与他的团队总是以最大限度的敬意和最大程度的努力来对待每一个项目，每一个工程，我深感其不易。他们殚精竭虑，夜不能寐，从立意上、形式上、细节上都苛求完美，这正是当代中国建筑设计领域所应当学习的。因此，他们的成功，当属必然。

从近几十年建筑历史发展的经验来看，自现代主义以后我们有很多各式各样的“主义”，但它们存在的时间都很短，稍纵即逝，并没有形成一定的、为学界所普遍公认和接受的理论和实践。因此，长期以来，仁者见仁、智者见智、莫衷一是。中国的建筑设计也进入了一个激变的时期，社会的迅速发展带来机遇、荣誉与挑战，而就在这样独特的背景下，罗凯另辟蹊径，在“历史•文化•传承”的主题下开始了他的探索。

从“华东政法学院”开始，经“复旦”，后“水产”，而后“应用技术学院”，件件作品都可以让我们细细品味。罗凯在“古典”与“现代”中不断平衡，在“思想”与“市场”间反复较量，在与业主的磨合中，逐渐提出“新古典、现代感”的总构思，在建筑的细部上又加以认真细致的推敲，提出了“造新如旧”的构想，用这样的方法彰显文化底蕴，用探索赢得了市场和业主，收获了相当的成就。

罗凯沉浸、研磨古典历史人文建筑有年，当然也包括其他各种方向的探索，时间在前进、事物在发展，人们的认识、思想、看法也是会不断变化发展的。如何把握住这一时代脉搏，以期做到与时俱进，仍然是放在罗凯和我们所有建筑师面前一个永续不断的课题。

探索、研究是很艰巨的事，也是非常有意义的事，又是绝对有必要的事。六年的时间不算长，可也不算短，罗凯还很年轻，我们祝愿他在已经取得大量成就的基础上，勇往直前，更要不断思考，进一步丰富、充实自己，我相信罗凯一定会做出更大的成绩，取得更大的成功！

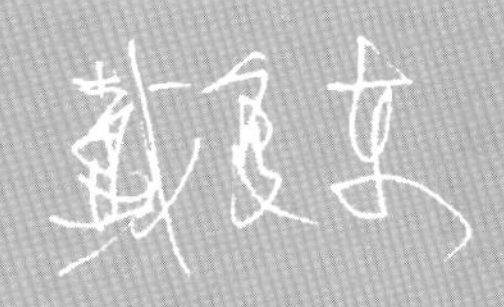

中国工程院院士
同济大学建筑与城市规划学院名誉院长
2006年国庆日

Preface

Under the leadership of Mr. Larkin, DHGP has finished eighteen cultural and educational construction projects for Shanghai universities since 2001. The book *Culture. History. Inheritance—Masterpiece of DHGP Cultural&educational Architectural Design* will be published by China Architecture and building Press. As one of my graduated students, Mr. Larkin wanted me to preface the book. His hearty invitation was so hard to resist that I accepted it deferentially.

I feel deeply touched on seeing the rich materials of this aplenty collection. Firstly, in such a short time, DHGP received a great number of projects and occupied half of the whole construction market of Shanghai universities. They have won their own glory. Secondly, Larkin and his team strive for nothing short of devoir and effort to each project ranging from master plan to single design, from interior decoration to exterior environment. They racked their brains and worked late to bed, seeking the perfect effect of ideas, forms and details. What they did is exactly what we lack in the designing field in China at the present stage. Therefore, they are deserved to succeed.

Looking through the experience of the architecture history within decades of years, it is easy to find that after Modernism there are various "isms", which are like flashes in the pan. Obviously, there is no certain theory and practice recognized and accepted by academic field. Therefore, the benevolent see benevolence and the wise see wisdom; it's entirely a matter of preference. Chinese architectural design is also coming into a new period that dramatically changes. The rapid development of the society brings opportunities, credits and challenges. Under such a special background, Mr. Larkin found another way that he started his exploration under the theme of humanity, history and inheritance.

Beginning from 'East China University of Politics and Law', through 'Fudan University', 'Shanghai Fisheries University', and then 'Shanghai Institute of Technology', We can slowly taste and repeatedly enjoy each of the works which is like having a cup of coffee. The aroma stays in your teeth for quite a long time. By continually balancing 'classicality' and 'modern', by constantly stuggling between concept and maket, by furthering mutual understanding with clients, Mr. Larkin gradually brought forward the whole concept of "neoclassic, sense of modernization" .Then he carefully thrashed the details of buildings and brought forward the idea of 'build a new construction but looks like with a long history'. He displays the charming of culture by using this method and receives recognition from people and public voice, gaining considerable achievements.

Mr. Larkin is immersed in studying classical, historic and cultural architecture for years except for other aspects. The world is developing with the lapse of time, so are the understanding, ideology and perspective of people. How to conduct exploration and research to keep pace with the time while focusing on the change of the world is still an enduring and eternal subject in front of Mr. Larkin and other architects.

The words above, compared with Larkin's six years' exploration, is just like a drop in the bucket or a skimming over the surface. Exploration and study is arduous but also meaningful and necessary. It is also a must for achieving success. Six years is neither long nor short. Larkin is still young. Based on the success he has already achieved, we wish that he will not be self-content but blade new trails, keep thinking and take different advices to complete himself. I am firmly convinced that he will have more success and get great achievements.

Fudong Dai
National Day in 2006

古典与现代的结合，东方与西方的沟通，学术与艺术的交融，让建筑成为倾诉美的音符。这就是建筑设计师罗凯的追求。

华东政法学院院长 何勤华

百年复旦，桃李满天下。过去的学子和今天的学子记忆中最美好的印象，除了师生情和同学谊之外，就是复旦的校园和建筑。校园和建筑既是复旦过去百年辉煌历史的见证者，也是未来百年发展的见证者。因此，当我们在规划新校区的时候，我们必须严肃地考虑一个问题，那就是如何体现历史的传承和未来的创造——这是一种使命，更是一种情结。所幸由罗凯先生担任总建筑师的团队，在复旦江湾新校区的规划设计中，以深厚的文化底蕴和缜密的专业逻辑，通过建筑语言，充分展现了复旦的人文精神以及向世界一流大学迈进的崭新姿态。

复旦大学常务副校长 张一华

在声势浩大的大学造园运动中，也造就和培育了一批优秀的建筑设计大师，罗凯先生就是其中的一位。我在学校新校区的设计中有幸结识了这样一位年轻有为的设计大师。他徜徉在高校建筑设计的校园里，善于站在巨人的肩膀上，注重各学校文化和历史文脉的传承，同时又大胆地吸收西方建筑设计的精华，创造出个性鲜明、高度折衷的“新历史主义”的设计理念，形成了独特的建筑设计风格。在设计的作品中展现出了一种平衡的美感和协调的真实。

上海应用技术学院党委书记 祁学银

上海工程技术大学松江新校区的建筑群像一幅马蒂斯的构成主义的绘画，更像是一首讴歌科技之美的理性主义乐章，罗凯先生设计的“现代工业训练中心”和“体育馆”就是这幅绘画的亮点，是这首美妙乐章中的和谐音符。华谏国际的建筑师们，以他们高超的艺术和技术才华，为上海工程技术大学的校园注入了一种科技精神，这种精神支撑着我们共同创造一所现代化大学。

上海工程技术大学校长 汪泓

在华东师范大学闵行校区的建设中，我认识了作者，在不断的争辩中我们相互有了了解并结下友谊。这是一位非常有个性有才华的青年建筑设计师。他对中西方建筑文化的认识和理解有着超常的天赋，读了本书多少可以感受这一点，这为他的设计项目和创作作品增添了不少耐人寻味的色彩。他不仅是在创作，更是通过作品表达了他的创作思想。

我很佩服他能在很短的时间里感悟到一个学校的文化，并在他的作品中把它表现出来。他那种追求作品完美的执著精神是很让人感动的。一次偶然的事件，我的同事告诉我：每当我们争论之后，他都会“猫”在现场或校园某个角落，静静地思索数个小时。第二天他又会滔滔不绝地叙述他的新方案直到你接受他为止，当你再去仔细品味这个作品时，你会惊奇地发现它还是来源于作者的原创思想，只是它变得更完美了。闵行校区学术交流中心的设计就是一个例子，他足足修改了十二稿。他把建筑看成是一种创作，我非常喜欢他这一点。

我认识很多建筑师，作者是其中很出色的一位，他们都是我们国家建筑界的未来，我多么希望我们的建筑师能坚持不懈努力，形成有中国特色的建筑流派，留下更多的传世之作。

华东师范大学副校长 [illegible]

百年沧桑，弹指即过。每当清晨漫步在上海理工大学的老校园时，薄雾中的老建筑都会给你讲述学校百年历史的风风雨雨，它们已经沉淀为学校的文化，成为了这个城市的记忆，同时还将继续伴随我们开创新的篇章。

罗凯先生，一位执着的历史主义建筑师，在上海理工大学新教学楼的设计中，通过其敏锐的思维，运用娴熟的建筑语汇，表达了他对历史的解读。在尊重历史的同时，建筑本身也体现了上海理工大学的人文精神和在新世纪继往开来，迈步国际著名大学的决心和勇气，并给上海理工大学的百年校庆奉献了一份珍贵的礼物。

上海理工大学副校长 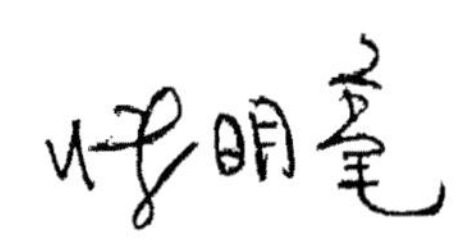

一代又一代莘莘学子走出了校门，校园里留下了什么？一处又一处大学新校区建立起来，校园向人们展示了什么？罗凯的校园规划设计，既延续了高校的渊源和文脉，又体现了现代的浪漫与理性；既反映了高校的科学精神和人文关怀，又彰显了各自的特色与个性。罗凯的作品和理念与大学的神和韵得到了完美的结合，我们有理由期待电力学院又成为上海高校造园新地标。

上海电力学院副院长

打造中国现代高校建筑史第七节点的雄心，从中我们可以看到，罗凯先生对当今“高校造园运动”中表现出来的急功近利、心浮气躁现象的忧虑。这种雄心和忧虑，展示给广大读者的是年轻设计师的良心、责任、理想与追求。

凭借着设计师对东西方建筑历史和文化的深刻感悟，对高校校园规划和建筑特质的把握和对每个建筑细节的孜孜以求，罗凯先生的作品已经在规划建筑设计和建成效果上独树一帜。我们怀着十分欣喜的心情见证着其“新历史主义建筑风格”在高校的实践，分享其在教育建筑设计领域成功的快乐。

上海水产大学副校长

时间是检验一切事物的最好方式。宝山气象中心算来建成已近三年，其简洁高雅的气质依然令人倾心。我一直觉得“恰如其分，精致典雅”最适合用来评价这个精美的建筑。宝山气象局的新建筑也成为气象系统公认的亮点和景观。而它的创造者，罗凯先生同样具有这种“恰如其分”的气质。并且他将这种气质投入到工作、生活、对社会责任的践行以及建筑创作之中。也许这就是君子之道吧。

上海宝山区气象局局长

建筑应该是社会发展的载体。

建筑应该反映社会历史文化的传承。

建筑应该体现当代民众的审美情趣。

艺术实验中心是中国艺术类院校中首个功能齐全，产、学、研一体的综合性教学实验平台。在中国艺术类教学的现代化进程中写上了浓重的一笔。

而艺术实验中心设计的本身同样是一个涅槃的过程。华谏国际的建筑师运用和谐的比例和尺度，能够净化心灵的空间特质，来体现对功能的尊重和理解，力图奉献出一个艺术的“雅典学院”。他们的作品就像是平实优雅的散文一样，在返朴归真的宁静中，为我们娓娓道来了那个寻找艺术真谛的恒久故事。

原复旦大学上海视觉艺术学院副院长

历史 • 文化 • 传承

罗凯 自序

引 子

封存的记忆就像进入泥土的果核，在不知不觉中就长成了一棵大树。

作为建筑师，作为正值创作之年的建筑师，作为饱受中西方文化洗礼，又生存在我们这个年代的建筑师，我行将度过自己的第三个本命年。回首这些年的建筑创作，清灯黄卷、涕泗长流，在林林总总的项目中，找寻着自己心中奔泻的激情和畅想，也承受着期待的心情和信赖的目光。岁月流逝，年龄渐长，自己的要求与别人的要求，让我愈加领悟到自身背负的义务与责任，这种来自外界同时又来自于内心的力量，使我一次又一次去发掘想要表达真实情感的东西，力图在作品中展示出一种平衡的美感和协调的真实。

只是，我还需正视我们生存的这个时代。简单来说，哲学也好，艺术也好，都是对现象的猜测或拙劣的仿制，这个世界本身有着它自己的逻辑。我们的思想似乎处于一个混合的时代，就像各种染料搅混在一只盛满清水的缸里，你从中已照不出自己的影子，衡量建筑好与坏，甚至善与恶、美与丑的标准都已渐渐模糊。在这个时代，也许总有类似的场景发生：你正在享受着清新的晨风与树丛中的鸟鸣，却不得不同时忍受着从邻近化工厂飘来的烂苹果味和氧化铁厂上空吹来的甜风。目前的现实是：一方面，我们面临着普世价值观与评判体系的缺失；另一方面，作为社会精神支柱的知识分子却对它熟视无睹。建筑师理当是知识分子的一部分，如果要在这个时代缔造精品，缔造流芳百世的作品，背负历史责任感的建筑师无疑还任重而道远。

在我们这个时代，科学、艺术等领域所表现出来的创造性确实难以掩饰其内在的贫乏。一个时代总是将它的特征凸显在哲学与艺术里，沿着这条线索，也许可以探查到我们这个时代的贫困本质：先是思想的贫困，再是情感的贫困，然后经由艺术世界与外在生活将其传达出来，这也许是生活在这个时代的建筑师建筑创作的真正困惑所在。

帕斯卡尔说，人是一根会思想的芦苇，我们的全部尊严就在于思想。可我们究竟要思想什么呢？我们的思想如何才能与我们的现实达成一致？我们可以做一些有益的工作吗？庄子在他的寓言中提到了一种鸟，“非练果不食，非醴泉不饮”，新千年的中国建筑师果真能如此对待自己的设计吗？——在这个混合的时代，建筑师必须为自己的灵魂制订规则。夜深人静处，我有时不免惆怅，觉得要能感悟世间真谛真的很难。面对思绪的驰骋和案头堆积的资料，有时唏嘘扼腕，有时思绪澎湃，我就像一个炼金术士那样小心翼翼地雕琢着每一个建筑的细节，谋划着淤积在心中的梦想。或许我该做些什么，留下些什么，或许这些只是建筑历史的一瞬，这些作品和思想或许会像耀眼的流星划过，或许会在史册中有一点点沉淀，于是，就有了这本作品专辑。

这些设计作品饱蘸了对历史与文化眷恋的激情，凝结着对文化教育建筑的感悟和思辩，这是我内心辐射式的历史道白，也是我与中国高校造园史的超然对话。但愿它唤起的不单有苍凉而执拗的使命感，也有对命途多蹇的建筑思潮悲欣交集的感戴与超迈，正如一位感悟特异的先知所云：“夜正长，路也正长”。

一、从问题开始

这几年，我有幸投身于上海的高校造园运动中，先后主持设计并实施了十余所上海高校的新校区，承担了从校园规划到建筑单体、从室内设计到绿化景观的总体方案设计和施工图设计的全过程，也参与过一些高校规划及建筑设计方案的评选工作，身感重任在肩。

当我与主管校长、基建负责人交流时总是有很多很多相似的问题：

● 建筑造型选择古典主义风格还是现代主义风格？

● 为什么规划中总要有一条中轴大道通向图书馆或行政楼？

● 规划设计师都强调功能分区，人车分流，但教学区与生活区有1~2km的距离，很不方便，能不能将教学、餐饮、生活区混合设置呢？

● 采用通高二层的大台阶作主入口很有气势，但使用上很不方便，前几天下雪，大台阶上不少人摔跤；校门主入口是“门面”，是重中之重，若主校门有横梁似乎不妥，一边一个门卫已成为上海高校校门的固定模式。那么，校门做多高和多宽才合适呢？校门不大没有气势，校门大了，一般都有300多m^2的建筑面积，过于浪费，美观和实用之间该如何取舍？

● 学校里新建的十几栋建筑是几个不同的设计单位设计的，建筑风格有很大差异，应该是统一风格还是各具特色呢？各个单体的色彩选择是强调统一还是差别各异？若有差别，色相相差多少合适呢？面砖是采用光面、毛面还是亚光面呢？单看一块面砖判断色彩，容易偏差，一个平方米的试贴与完整大面积的色彩还是差别很大，面砖色彩该如何选择才会有新意呢？

● 采光中庭空间是建筑师常用的手法，采光顶好是好，只是能耗太大，采光顶跨度大，清洁难度高，有没有其他空间营造的手法？

● 新建校园中都有人工开挖的或大或小的湖面，是做亲水驳岸、自然驳岸、硬质驳岸、木栈道还是卵石滩呢？抑或都做一点，怎么权衡其比重呢？水体自洁系统是否行得通，水不会发臭吧？

● 外墙需要强调质感，是用真石漆还是干挂石材呢？干挂石材是用镜面石材还是烧毛石材？是用凿毛石材、剁斧面还是荔枝面呢？石材干挂太贵，湿贴又渗水渍，怎么才能既美观又便宜呢？

● 窗框是氟碳烤漆还是粉末喷涂？选择什么颜色？是采用镀膜玻璃还是Low-E玻璃？其透光率和反射率参数如何确定？

● 校园环路是种梧桐还是香樟？种落叶树还是常绿树？用杀头树还是全冠树？加拿利海藻、布迪椰子、华棕这么漂亮的树种在上海这地方就是难活，园林公司说包活两年，但两年以后死了怎么办呢？

● 桥栏杆用什么材料最合适？是木材、仿木、混凝土、钢还是铸铜？桥栏杆是简单一点还是复杂一点呢？桥的效果图看看还可以，造出来的效果为何相去甚远呢？

等等，等等……

这些问题，书上没有，建筑学的老师也没教过。一旦深入项目，这些问题便会扑面而来，接踵而至，而这也正是中国高校造园运动的普遍问题，也正是我们这个时代暴风骤雨大规模建设和千篇一律的设计规划的真实写照。要在短时间内缔造一所新型大学，要在规划设计和建成效果上独树一帜，在短时间内制造出与众不同、流芳百世、成本低廉、好看耐用的高校建筑，对建筑师来说，既是心灵的涤荡和涅槃，同时也是难以完成的艰巨任务。

单独研究上述问题个案，似乎都有解决方案，但问题越来越多，越来越杂，就陷入“头痛医头，脚痛医脚”的尴尬境地，有没有举纲张目的宏观思考方式呢？有没有营造现阶段大学校园真正美与丑的判断标尺呢？螳臂挡车，我试图解开这个死结。

二、“历史”的过去与未来

华谏国际（DHGP）真正在高校建筑这块领域中有所作为，凭借的是上海松江大学城华东政法学院（以下昵称“华政”）这一项目。作为在中国常规建筑学教育与现代主义神话熏陶下培养出的建筑师，要一开始便主动设计出现在“华政”样貌的建筑，那几乎是不可能的。所以当时的中标方案与现在已建成的面貌相距甚远，以至于后来从方案到施工图的过程中，对建筑单体的修改频率之大、时间之长、任务之紧，在华谏的设计史中也是少见的。犹记得华政的教学楼是画一层施工一层的典型双边项目，而图书馆的塔楼更是历经了十数稿的修改，光这些修改稿也都可以集结成书了（图01，图02）。

“华政”从方案到施工图的修改过程，是一个设计师从理想回归现实的修正过程，是一个由高高在上、自我傲慢，向主流民意与大众审美趋向的过程。没有经历过这个过程的设计师，很难理解我们对所谓“新古典”从排斥到接受再到推陈出新的心理转变。而时下的大部分所谓新锐建筑师，仍然在自我把玩着先锋的词汇和语句，以提升与引导大众的审美情趣为己任；或者

01 02

重复着面孔苍白、情感匮乏的建筑表情，以满足个人或所谓的社会正统价值。实际上，中国的建筑学教育几十年来有意无意地在神话某种东西，而又丑化了某种东西；而在我看来，历史的前进过程中，无论哪类风格都有其存在的必然，而没有高下之分。历史总在浪漫与理性中反复，这与社会背景和人文发展密不可分、息息相关。建筑这门学科也不存在绝对的正确与错误，只存在于极致与非极致的区别。

整个五千年的中国历史关于建筑的专著有两部是建筑师必读的：宋代李诫的《营造法式》和明代计成的《园冶》，中国也许以前就没有建筑师，鲁班也只能叫做建筑匠吧。当今的中国建筑直接从大屋顶奔向了现代主义，经过改革开放的若干年，现代、后现代、折衷、古典、解构主义扑面而来，建筑师、业主、甲方都被弄得眼花缭乱、一头雾水。中国现阶段的建筑设计既可以是解构主义的，也可以是晚期现代的，既可以是超现实主义的，又可以是折衷主义的，不一而足。因为中国建筑既没有经过哥特式、文艺复兴、巴洛克、洛可可，也没有新艺术运动，历史的断层造就了今天百花齐放、百家争鸣的时代，也造就了价值观评判体系的缺失，建筑创作繁荣的同时也存在着评判盲点和认识误区。

同一风格的东西，我们往往能一眼看出其好坏，是否用心、是否细致、比例尺度推敲是否到位，这很容易判断。但不同风格的东西往往很难用同一套价值体系来评价好坏，因为她们是不能类比的东西，只能说哪一种风格可能更合适校园建筑这一形态。每个校长拿着一块地，如同拿着一块布料，是做成中山装呢还是西装？是牛仔裤、吊带衫抑或是旗袍？然而，喜欢以“主义”归类的建筑理论界偏偏要在牛仔裤与旗袍之间分出高下，用学术的清高拒绝实际的操作，用理性的观点围堵感情的宣泄。

在我眼里，西方建筑史从古至今一直在技术与浪漫、理性与感性之间徘徊，不断反复，发展至今。最近一次理性向感性的摇摆是发生在开始反思现代主义的片面与乏味时，以文丘

里为首的“保守的反现代主义”，提出了“历史主义”和“民间艺术”是发展当代建筑的两只划船的桨，主张建筑创作“一手伸向古代，一手伸向大众”。到了20世纪中后期，后现代主义的产生再一次掀起了使用历史片段来反对现代主义建筑的创作之风。历史主义作为后现代主义的一个部分，在西方出现有它的合理性，它从人文主义立场出发，对现代化过程中的人性、价值观和意义等方面提出了质疑，并充满了文化上的忧虑。

基于这种开放的历史观，我们可以清醒地认识到，任何建筑思潮的产生和发展都不仅仅是建筑领域自身的产物，也不仅仅是任何天才大师所决定的，而是社会、文化、科学、思想等综合作用的结果。建筑思潮不过是历史大潮中的一朵浪花，如果我们剖开任何一个历史横断面的话，我们不难发现，任何形式的建筑表现都存在着深刻的社会根源和时代烙印。

上海高校造园史乃至于中国现代建筑史都与西方建筑史有着千丝万缕的联系，我们可以在史学比鉴中寻求佐证，探询现阶段中国高校的主流设计风格。而进入新千年后，大学校区建设的决策者也自发地对新校区建筑刻板、生硬、千篇一律的问题提出质疑，主张个性化和重视建筑环境及文化氛围的营造，对规划建筑方案的选择反映出他们在“现代”与“传统”的冲突中困惑徘徊、复杂微妙的心情。

我们可以看到力扛历史主义大旗的“地产大鳄”——珠江国际、合生创展、浙江绿城，之所以在众多的房地产企业中脱颖而出，靠的是对历史主义建筑设计的不懈追求。而他们之所以在众多的建筑风格与样式中选择历史主义，当然是出于开发商的战略谋划和市场把握，更深层次地反映出在当前的社会环境和经济状况下国内消费群体的审美倾向。

我们对高校建筑形式的宏观发展方向从大历史观的角度做了一个清晰的判断——正如台湾学者龙应台所说：从传统走向现代的过程中，所有努力都往“现代化”的方向走去，但发展的目标其实不应该在“现代”的那一端，而是在传统与现代这两点中间，就如同“双面苏绣”，一面是科技、一面是人文，协调地合为一体。

三、校园的精神与气质

人们常常由于技术获得了“幸福”的同时，却失去了对“崇高”的追求。我们不仅仅是“存在”，而是要“诗意的栖居”。盲从于现代性，就没有意识到感情的力量、社会环境的力量和历史传统的力量，而这些值得强调的力量构成了一个大学校园的核心精神。

现代主义理论是建立在对建筑物质文化的诠释基础之上的，其价值判断具有鲜明的条理化倾向；历史主义则以建筑的精神文化内涵为其理论核心，价值判断展示出浓烈的人文关怀与人本思想。在2001年之前，上海的新校园建筑采用现代主义的手法，大量性地建造了一批批量式、流水线产品，已基本解决了使用、功能、安全的问题，但千篇一律的形式和色彩已经让我们倍感厌倦。对校园建筑而言，人文层次是高于技术层次的更高追求，当我们现实中的功能、交通、设备等等问题都被解决之后，我们需要进一步考虑如何塑造校园环境的气质，这往往是现代主义建筑所难以企及的高度。

华政的成功使我们敏锐地捕捉到当今大学校园所需的精神与气质 ，我们在跌跌撞撞中前行，华政的经验在接下来的几个设计项目中延续……

复旦大学新江湾城校区，从其本身的概念来讲，就是一个文脉延续的问题，是把老校区的情感移植到新校区的问题，什么样的建筑形式才可以承载“百年”这一称谓？当我们在复旦相辉堂前的大草坪上、上理工的大礼堂前抑或是华政思孟堂旁漫步时，我们会惊讶于这种历经百年、历久弥新的魅力，它们也曾被修缮，但只是力图恢复原状而已。复旦江湾新校区建筑群三段式的比例，精巧的细部、拱券、回廊是在历史与现实、建筑与环境之间建立了一种文脉上的联系，并产生了强烈的修辞效果，具备了在时间流逝中始终经典的价值。你可以赞美它的深厚，也可以批评它的古典；你可以抨击它的孤傲，但你无法否认它所具有的深厚的文化积淀和动人心魄的精神依恋。

上海应用技术学院漕宝路南校区新图书馆的设计强调新建建筑与周边环境、传统历史和谐共生，当新来者问道："新图书馆在哪里"时，城市设计与建筑设计的最高境界已经达到——"造新如旧、宛若天成"，这才应该是大学校园新建建筑历史观的最高体现（图03，图04）。

歌德说过："建筑是凝固的音乐"。历史主义建筑的大气恢宏让人心动景仰，它的博大精深在不知不觉之中便把人引领进一种平和而又华丽的境界，建筑的本身就蕴含着无与伦比的韵律，激昂和柔和的反差正体现着深沉的内涵。更重要的是，历史主义建筑的精髓在于表现人类的思想，或是内心深处的东西，并不是单纯的渲泄。漫步在华政的老校园，浅色石料和

03

04

暗红色的砖墙在斑驳的树影之中默默无言，让人安静内敛，感怀深思。

气质与精神对于一所大学校园意味着什么？校园总体环境作为一种符号系统，蕴涵着复杂多样的意义，传播着丰富的历史文化内涵，镌刻着特定历史时期学校发展的轨迹，体现了一个学校蕴涵的教育传统和文化积淀。建筑包含着必须考虑的所有因素——不仅是实用的、技术的和经济的因素，而且也包含心理的、情感的、美学的、精神的因素，不是一种纯客观的事件，也不是纯主观的意识，而是主客体的交融与结合。为什么我们一想到剑桥、耶鲁的建筑就激动不已，因为当我们看到它，不仅仅感受到的是建筑的"美"，还有校园的历史、文化、氛围等种种载体，折射出的是校园历经百年逐渐完善的文化内涵和历史印记。具有历史感的校园，告诉人们往日的辉煌，任何人都能够从中想像或感受到自豪、骄傲、崇高、不懈的追求、等等复杂的情感。用这样的观点而非个人的好恶爱憎去看待和审视建筑风格，也许正是在浮躁之风日盛的今天，决策者和建筑师所应有的心态。

四、细节的创造与价值

对于一幢建筑、一个单体来说，往往是细节决定成败。无论是现代主义还是历史主义，都有好与不好、美和不美的作品，即便是"少即是多"的现代主义作品，玻璃与钢的节点、显框与隐框的分割，都是注重细节的表现。缺乏细节的现代主义是粗陋的现代主义，缺乏细节的历史主义是庸俗的历史主义。细节的塑造取决于建筑师的职业道德与素养，也取决于甲方的资金投入与控制能力。如何能用最少的钱塑造充实完美的建筑，获得性价比的最大成功，这是一个难题。

我们所谈论的细节，不仅仅是效果图上的细节，更为重要的是在施工图上所能表达的细节，概预算中可以控制的细节，而决策者对设计师的要求经常是：不怕做不到，就怕想不到，但在众多花花绿绿的效果图一遍一遍地审视之后，却很少认真地研读施工图（研读墙身大样、

研读外墙分割、研读钢构节点）。效果图很精彩，造出来大模样也差不多，但少了很多东西。少了什么？细节！人可以体会的细节，接近人体尺度、可以触摸、可以感知的细节（图05）！

为什么会这样？在效果图阶段，设计单位可以暂时抛开资金的束缚而大胆创新，但到了施工图阶段，中空玻璃、氟碳钢构、点式幕墙等的昂贵造价成为甲方不可承受之重。于是减去节点、减去细部。为什么都是幕墙、都是铝板，德国造可以那么浪漫、轻盈、纯净，中国造却是如此平庸、粗糙、浑浊？关键在于许多人都忽略了：某些细节的表达并非是现阶段中国大学的财力所能负担的。拿服装作例子，没有很好的料子，做工很粗糙，又追求奇奇怪怪的东西，这种服装只能是地摊货；而在现阶段的经济技术条件和施工水平下，追求所谓新奇独特，只能是建筑设计的地摊货；高校建设经费拮据，单方造价控制在2500~3000元，追求所谓的横空出世、标新立异，只能成为高校建筑的地摊货！看一下以下几所大学的造价：

华东政法学院教学楼：2500元/m^2

上海工程技术大学现代工业训练中心：2300元/m^2

上海应用技术学院南校区图书馆：3300元/m^2

我们的2500元/m^2同样要解决流芳百世的问题，建筑造价决定了其本身运用的设计手法和建筑形式：抛开大面积的玻璃，大体量的钢构，不要被形形色色的图面效果，甚至是当前技术水平所无法达到的效果所迷惑，活用、用好砖和混凝土来打造我们所要的一切文化和精神的本质，探询在既定造价下的平实设计手法，表征高校建筑的文化性和历史感，营造近人尺度和文脉传承。

真正好的设计，应当是一贯的、实际与效果图一致的甚至是更为出色的设计。我们可以看到，华谏国际（DHGP）"后历史主义"大学校园的作品，仅仅采用混凝土框架和砖砌的墙身，就可以使墙面进退凹凸有致，表达出耐人寻味的细部，把优雅的历史主义符号和潇洒、精致的现代手法融为一体，以节奏的强烈和形式的清晰，体现出凝练而富有个性的主题及富于想像的建筑空间。在为数不多的华谏国际（DHGP）现代主义作品里同样还是运用砖和混凝土来打造细节，外墙少用幕墙、铝板，而是巧妙运用面砖、涂料，造价低廉、精巧雅致（图06，图07）。

06

五、华谏的历史与方向

华谏国际（DHGP）的设计发展历史，在过去的六年间，呈现出多姿多彩的面貌：上海应用技术学院南校区是一种中国20世纪五六十年代深受苏联东欧式建筑影响的所谓"中国固有形式"风格，雄浑坚实；而华政新校区则打上了西欧哥特式的深深烙印，直插云霄；复旦与华政又有所不同，平顶方塔代替了刺破天空的尖塔，她是纯朴希腊式风格的延伸，承载着海纳百川、跨越民族界限、追求至真至善至美的包容；而上海水产大学则如同根植于中国文化的水墨画，既

07

典雅端庄，又气韵灵动；上海电力学院是一种现代主义雕塑感的体现，用理性的黑白灰三色勾勒出学科特征。如此多姿多彩的变化不但反映建筑师的功力，也正说明了华谏的多变创新特征。其中，历史主义倾向的项目犹如最闪亮的明珠，呈现出华谏国际（DHGP）“后历史主义”三部曲的发展道路。

08

上海松江大学城华东政法学院这个项目，在华谏的历史上是具有里程碑意义的，当时国内简单粗陋的复古风格不乏其例，甚至一度到泛滥成灾的地步。而华政新校区所采用的历史风格实际上正在脱离纯粹的欧式古典，而根据现代的工艺水平与审美品味有了大幅度改进，可以说是一种现代古典或新历史主义。它一方面吸收了经典欧式建筑的比例，并截取那些有明显特征的符号，甚至哥特建筑特有的夸张具表现力的符号；另一方面它把古典的符号与现代的材料和工艺相结合，创造一种全新的魅力，是在现代工业的基础上表现传统手工的美（图08）。

其后是上海应用技术学院漕宝路南校区的图书馆与艺术楼，与华谏的其他项目相比，这个项目只有两栋楼，相对较小，但是早期项目中最为经典的一个：洗尽铅华、低调质朴，在细节的把握上更为大巧不工；她秉承的是中国20世纪五六十年代“中国固有形式”时期的气质，更绝妙的是在一片老建筑中，他们“造新如旧、宛若天开”，令人感受不到一丝做作的痕迹，此为上品（图09，图10）。

09

10

复旦大学的江湾校区更无需赘述，气势宏大，不过令人纳闷的是，不知为何总有人说她与华政是一个风格，而在我眼里她们不像至极。且不用说色彩上的差异，一个具有法学的森严气象，哥特式的扶壁与塔楼直插云霄；一个具有古希腊罗马的宏大胸怀，平头方脑，敦厚坚实。想来想去，终于明白，原来世人眼中的古典只有一种，即“非现代”，殊不知，古典历史主义与现代主义一样，也是有不同性格、不同气质的多种表达，有的气势雄浑，有的浓厚质朴，有的典雅俊秀，有的意境深远，需要人们慢慢品味而不是遽下结论，在推崇古典的人眼中，现代主义的东西又何尝不是千篇一律呢（图11）。

我们谈到了华东政法学院、复旦大学、上海应用技术学院南校区，以上三个项目是华谏国际(DHGP)“后历史主义”三部曲的序曲。这个时期，在细节设计手法上，几经磨砺愈见成熟，古典的片段、语句与符号开始运用自如；在造价控制上，我们对相对低成本与低造价的条件下建成优秀作品有了更多信心，事实证明，用砖与混凝土来打

11

12

造细节，性价比确实很高。

但是有人不禁会问，在华谏的新设计中，总不能再造一个华政或是复旦。是的，我深深懂得没有历史印记的校园无异于文化的沙漠，而每个校园的历史印记都不能等同，我是一个历史主义者，在成就了华政、复旦、工技大等大学之后，我陷入了深深的思考，设计师不安分的内心蠢蠢欲动，开始觉得以前的道路略有保守，想开拓一条新的道路，一条不那么忠实于某一类风格，不拘泥于某一种手法的道路。

机会来自于"上海音乐学院"，这个投标方案是华谏国际(DHGP)"后历史主义"风格的转折点。在几栋单体中，我们可以看到力图用玻璃和钢构来表达古典的方式，也可以看到早期的古典成熟风格。遗憾的是，由于种种原因没有中标，更进一步的探索嘎然而止。上音为我们打开了一道门，华谏进入了三部曲的第二阶段，此时，上海水产大学临港新校区与上海应用技术学院奉贤新校区的项目几乎同时展开（图12）。

上海水产大学的单体设计中，我们以简练质朴的建筑语言体现高校建筑舒展大气的开放性和文化性，色彩以黑白灰作为校园基调，体现含蓄内敛、古朴淡雅的校园气氛。建筑细节低调统一、愈久弥新，因此，在选材上摒弃铝板、彩色道砖之类的材料，而用粗石地坪、块石墙面、青灰色面砖、黑色压条、素钢与玻璃的交织……随着时光的流逝，岁月将印证她的永恒（图13）！

13

如果说上海水产大学临港新校区是一幅酣畅淋漓的水墨画，那上海应用技术学院奉贤新校区就是浓墨重彩的油画。立面设计思想强调中国传统与西方文化的结合，采用传统三段式的现代表达；立面色彩采用明度不一、多层次的暖色系，讲求校园整体形象的色彩微差；立面细节形成独具特色的"新应用学院符号系统"，既有一些中国式建筑的气韵，又有一些西方建筑的质感，还有一些后工业时代的模块，摒弃现代主义标榜的纯粹，追求新历史主义的建筑折衷美学（图14）。

上海音乐学院、上海水产大学临港新校区、上海应用技术学院奉贤校区，以上三个项目代表了华谏国际(DHGP)"后历史主义"的第二阶段，她们都不再追求某一类风格，却努力做到自成一格，自有一派，神似而形不似，这对于设计师是一个更大的挑战。

而我们对未来的展望就是华谏国际(DHGP)"后历史主义"的第三部曲：我们心中的方向

14

是更自由、更轻松、更活泼和更值得玩味的历史主义，也许是本书尚未收录的“山西运城学院”和“上海政法学院”（现代元素的比重更强，材质的古典表达），又或者是“上海音乐学院”中的一些尝试（钢与玻璃的古典符号，因为没有合适的项目，至今尚未实现）。

以上匆匆数言，于华谏国际（DHGP）几年来的不懈努力不过沧海一粟，回望历史，感慨万千；在建筑创作的道路上跋涉踟躇，才体味到知易行难；但只要心有方向，便可阔步前行。

总结

整个中国的建筑设计行业，其本身飞速发展所导致的急功近利正侵蚀着年轻设计师的灵魂，中国的建筑设计业所处的时代是产品时代远非作品时代，在一个前所未有的机遇到来的时候，中国的建筑界缺乏应有的准备。尚处于起步阶段的中国建筑设计界，其缺乏创造性是多方位的，这在理论研究和建筑评论上同样显露无遗。中国本土建筑师必须形成自己的价值判断体系和建筑理论导引，不断探询新兴的工程技术、材料知识、设计手段，以适应不断更新的市场环境，创造真正揭示人们内心情感的建筑表达方式，历史的重任无疑落在我们这代人肩上；而身处这个时代，更困难的事莫过于让建筑师明确以创作建筑精品为目标的历史使命；中国建筑师的现状，直接导致了中国建筑的平凡和空洞。同济大学伍江教授说：“中国建筑师在面临着令国外同行眼红的少有的时代机遇的同时，也面临着国外同行所无法想像的困境。他们在承担了让古往今来的建筑师都无法望其项背的巨大工程量的同时，却无法承担创造足以让后人骄傲的精神财富的重任。这不能不说是我们这个时代建筑师的悲剧。”

回到开篇的种种问题，提升到理论的高度，可以归纳成两个层次：宏观层次的方向把握和微观层次的细节处理。这两个层次同时分别反映优秀建筑设计的两类价值：空间创造的价值和细节刻画营造的价值。对于一个决策者来说，希望创造出什么样的大学新校园，为什么会如此难于选择？是因为社会普遍价值观体系的缺失同样殃及建筑设计业，但我们必须为自己的灵魂制订规则，必须给我们的内心设定价值衡量标尺。

我们不应被纷繁芜杂的问题所迷惑，不应为细枝末节而转移焦点，而应抓住教育建筑的本质。我心中呼之欲出的理想就是“营造具有人文情怀的历史主义大学校园”，这个理想也正在转化为现实的建筑作品巍然矗立在坚实的大地上。华谏国际（DHGP）的“后历史主义”风格——注重建筑文脉、推崇传统、刻画细节、在建筑中注入象征和隐喻的设计手法——的确摸准了市场需求和大众审美情趣的脉搏，使她的创作实践得到社会的接受和认同。

无庸讳言，老校园的历史印记已成为学校师生精神结构中的一种情结、一种埋藏在心灵深处的原形图式，只要遇到合适的环境和温煦的阳光，这颗文化与艺术的种子就会生根发芽。这也正是华谏国际（DHGP）校园规划建筑创作的原创动力和内在母题，也正是华谏国际（DHGP）“后历史主义”的现实溯源，是摒弃现代主义标榜的纯粹，追求新历史主义的建筑折衷美学，对比和协调在她所营造的肌肤表面跃动，近人尺度的细节饱满而真切。

我这些年对设计的最大感悟，正像台湾学者李敖所说的，“最贫乏而无益的，莫过于被现代性困惑纠缠，并且刻意忘记历史”。我们所需传达的讯息不应是描摹设计发展的历史，而是一种更为深切的领悟。的确，历史在化为时间流逝而去的同时，又化为一种神奇的有灵性的空间存在下来，我们大家要做的事情正是如此。

我们尊重历史，并不是因为借此可以怀古，是因为历史的线索提供在更高的水准上的理性的思考，如果我们对我们的校园寄托了生命的情感，就一定会被这些历史的线索打动，从中想像出校园的鲜活的精神整体。试想百年后，或许我们已经不在，但那些具有历史线索的校园建筑会告诉后人一切，告诉他们，我们确定的衡量标尺和我们所营造的细节已亘古流传！

图 示：

01 华东政法学院教学楼

02 华东政法学院图文信息中心

03 应用技术学院漕宝路南校区老教学楼

04 应用技术学院漕宝路南校区图文信息中心

05 复旦大学新江湾城校区行政会议中心柱廊细部

06 华东政法学院图文信息中心内院

07 华东政法学院图文信息中心柱廊

08 华东政法学院图文信息中心

09 应用技术学院漕宝路南校区图文信息中心

10 应用技术学院漕宝路南校区艺术楼

11 复旦大学新江湾城校区行政会议中心

12 上海音乐学院电化教学楼立面改造效果图

13 上海水产大学图书馆效果图

14 应用技术大学奉贤新校区教学楼效果图

15 上海音乐学院音乐会所效果图

16 华东政法学院图文信息中心塔楼

History • Culture • Inheritance

Larkin Authors Preface

Introduction

Like a seed of the fruit in the soil, memory sealed up for keeping grows into a tall tree reaching to the sky before we know it.

As an architect in the prime years of creation, an architect living in an age as ours, and an architect steeped in both Chinese and western culture, I am moving towards the third of my birth year (same with one of twelve animals representing the years in which people are born). A retrospective look at the architecture designs brought me back to the years of loneliness, hard work, and hardships. In the process of completing all these projects, I have been searching the passion and aspiration surging in me and receiving kind expectation and trust. As years elapse I grow older, and the requests, from others and from myself alike, stimulated me into an increasingly clear realization of the duty and responsibility on my shoulders. This power, from within as well as without, pushes me, time and time again, to explore whatever that can express my sincere feelings and emotions, and to reveal in my works a kind of balanced beauty and harmonious trueness.

Yet we do need to look squarely at the epoch in which we are living. Philosophy or art is, simply put, the conjecture about and the crude copy of phenomena of the universe, which has its own operation logic. Thoughts of our age seem to be in a state of confusion, like different dyes mixed in a vat of clean water, from which no reflection of yourself can be found. In an era as such, the standards for the evaluation of good or bad constructions, or even standards for good and bad, beauty and ugliness, are blurred. In an era as such, similar scenes are seen everywhere. When you are enjoying the fresh morning breeze and the twitter of birds from the woods, you have to endure at the same time the smell of rotten apple from the chemistry plant or the "sweet" wind from the ferric oxide factory in vicinity. The fact is that on the one hand we are faced with the want of value and evaluation system; on the other hand, the intellectuals as the backbone of our society turn a blind eye to such a shocking want. Architects are of course a special group of intellectuals. If architects want to create something immortal, those with strong sense of historical responsibility still have a lot to do.

The seeming creativity as seen in the fields like science and art in our era can hardly conceal its intrinsic tenuity. As a rule, an age reveals its characteristics in philosophy and arts. It is just from these two we might ascertain the root of this tenuity: first in thoughts, then in emotion, which betrays itself through both art and daily life. This might be the real confusion for the architects living in our age.

Pascal (1623-1662) told us that "Man is but a reed, the most feeble thing in nature, but he is a thinking reed. ... All our dignity then, consists in thought." But what should we think? How can we accord our thoughts with the reality? Can we do something beneficial to the world? Zhuang Zhou, a philosopher in ancient China, mentioned in one of his fables a kind of bird (phoenix) which "eats nothing but the fruit of the bamboo, drinks nothing but the purest spring water". Can the Chinese architects in the new millenary treat their own design in the same way? —— In such an age of confusion, we architects must set up rules for our own souls. Sometimes, in the dead of night, I could not help feeling disconsolate, because I find it is so hard to grasp the true meaning of the world. Galloping thoughts and the materials piled up on my desk leave me to sighs or surging thinking. I, like an alchemist, consider with great caution the details of every single construction, and design about the dreams accumulating in my heart. Perhaps I should do something, leave something to the world; perhaps this is but a transient moment of the history of construction. All my works and thoughts perhaps will flash away just like dazzling shooting stars, or they will add something to the history, hence this book.

These works, filled with the passion for history and culture, carrying my comprehension and thoughts, are my eradiating professions, and also a dialogue with China's history of university campus construction. I hope it will arouse not only the desolate and pertinacious sense of mission but also the gratitude to and transcendence of the trend of construction thoughts filled with misfortunes. As a prophet said, "The night is long, so is the our road. "

I. Questions to Start with

In the past several years, I am fortunate enough to have the opportunity to throw myself into university construction campaign in Shanghai. Successively I presided over the design and carried into execution of the new campuses for more than ten universities in Shanghai. I took in hand the design of campus layout, individual buildings, interior design, afforestation, shop drawing, and participated in the selection of public appraisal of capital construction. In this course I have deeply felt the weight on my shoulders.

When communicating with the vice presidents concerned or the officials in charge of capital construction I inevitably came across many similar questions:

● The modeling of the building should follow the classic style or the modern style?

● Why in layout planning is there always an axial main road in the middle of the campus leading to the library or the administration building?

● Layout designers always emphasize the divisions of zones according to their functions and the segregation of pedestrians and vehicles. Yet the 1 to 2 kilometers distance between the teaching area and dormitory proves a great inconvenience for the students. Can we mix the sub-areas for teaching, living and dinning?

● It adds magnificence to the building with grand steps leading to the second floor as the main entrance. But such entrance is very inconvenient in daily use: for example, many fell from the steps after the snow days ago. The main gate is "the gate of gates" for a university, the importance of which can not be overestimated. A horizontal beam might spoil the whole gate. A guard on each side of the gate has been stereotyped for universities in Shanghai. Then what should be the appropriate height and broad for a university gate? A small gate will not uphold the desired magnificence for the university, while too big a gate with a construction area of 300 square meters would be extravagant. Beauty or utility, which to choose?

● The ten-odd new buildings on the campus are from the hand of different designers, and the construction styles differ accordingly. Should we unify these styles or maintain the differences? The color of the individual buildings should be same or different? If different, how different should they be? The front tiles should be of gloss form, frosted form, or matt form? Judging the color by a single piece of front tile would result in inaccuracy. And the effect of one square meter trial paste of tiles would have a world of difference from a complete paste. How to be creative in selecting colors?

● Day lighting atrium is often employed by architects. A daylighting roof is certainly good, only it is very much energy-consuming. The big span of the daylighting roof will also add difficulty to the cleaning work. Is there any other method to add to the spaciousness beside the employment of daylighting atrium?

● Artificial lakes of all sizes are found on the new campuses. What kind of revetment is better for them? Hydrophile, natural, or hard revetment, wooden plank road or pebble beech? Or a bit of each? Then how to balance their proportion? Will the self-cleaning dispersal systems be workable? Would the water stink?

● For the outer wall, which demands an impression of solidness, which is better: to coat the wall with dry-hang stone slabs or to paint it with stone coating? What kind of stone slabs should be chosen, polished slabs, flamed slabs, chiseled slabs, axed slabs or brush-hammered slabs? Dry-hang is too expensive while wet-paste suffers water logging. How can we spend less but achieve more beauty?

● Should the window frame be coated with fluorocarbon baking varnish or powder spraying? What color would be good? Should we use coated-glass or Low-E glass? How to decide its parameter of transmissivity and reflectivity?

● The campus should be lined with phoenix trees or camphor trees, hardwood or evergreen, beheaded trees or full-crown trees? Such beautiful trees as Canary Phoenix dactylifera, Butiacapitata, and palm trees can hardly survive in Shanghai. Horticulture companies guarantee that the replanted trees will live for at least two years. What to do if they die after two years?

● What materials should best fit bridge balustrades? Wood, imitating wood, concrete, steel, or cast copper? Should the design of the bridge balustrades be complex or simple? Why the desired effect as presented in the architecture renderings is rarely found in the real building?

A list of such questions can go on and on.

The answer to these questions cannot be found in the textbooks, neither from the architecture professors. Yet once we submerge ourselves in the projects an overwhelming avalanche of such questions would come. These questions betray also the universal problems in the movement of campus construction in China, which are also the truthful reflections of the large scale hasty construction and stereotyped program in our age. To build up a new university in a short time, to be unique in programming and rendering effect, to produce beautiful, endurable university buildings at low cost but with unique and eternal charm, is both purification of the soul and nirvana and also a challenging task hard to complete for an architect. Taken individually, each of the above problems can be easily solved. But when the number and complexity of the problems increase, we will be trapped in an embarrassing situation of treating the symptoms but not the disease. Is there a macroscopical way of thinking as the Ariadne's thread that leads us through the Daedalus' labyrinth of campus construction? Are there any criteria to discriminate beauty from ugliness in campus construction? I write this book, as a mantis trying to stop a chariot, in the (perhaps vain) hope of unfastening this fast knot.

Ⅱ. History: Past and Future

DHGP distinguished itself in the field of campus construction by the project of East China University of Politics and Law (ECUPL) in the university town Songjiang District, Shanghai. It is almost impossible for an architect receiving both the regular education of architecture and the myth of modernism to produce an ECUPL as it is now. There are actually big differences between the design winning the bid and the buildings we see now. From the layout plan to the shop drawing, the frequency of revision, the length of construction, and the urgency of the task are rarely seen in DHGP's history. I still remember the construction of teaching buildings in East China University of Politics and Law, a typical bilateral project, in which shop drawing is finished just upon the construction of each floor.

The tower of the library has also undergone more than ten revisions, which might as well be collected as a book.

The course of the revising the layout plan and the shop drawing is also a course that an architect reverts from the ideal to the reality, a metamorphosis from an arrogant architect standing high above the masses to one inclining to the mainstream opinion and mass taste. Without experiencing such a course, an architect can hardly understand the psychological changes we have undergone in terms of our attitude towards the so-called New Classicalism, from rejection to acceptance and then to getting rid of the stale and bringing forth the fresh. However, some alleged avant-garde architects are either still indulged in the play of vanguard vocabulary, assuming as their task to guide and elevate the aesthetic taste of the masses or repeating the pale and emotionless constructions to satisfy their personal or the alleged legitimate social value. For several decades, the architecture education in China has been apotheosizing something and at the same time belittling something else. Yet to me, in the long course of history, each style has its own reason for existence, and no style is superior to another. History swings between romanticism and logos, a motion closely connected with the social backgrounds and human developments. Similarly, there is nothing absolutely right or wrong in architecture, with the only difference as perfection or non-perfection.

Of all the monographs on architecture written in the last five thousand years in China, two are must-reads for architects: The Craft of Gardens (1631) by Ji Cheng (1582-1642), and Manual on Architecture (1103) by Li Jie in the Song Dynasty (960£-1279). Perhaps China never had an architect in the real sense. Even Lu Ban (507BC£-444 BC, legendary founder of Chinese woodwork) can only be called a craftsman. Today, the Chinese architecture has rushed from vast roof to modernism. After years of reform and opening-up to the foreign countries, all styles of construction, European, modern, postmodern, eclectic, classical, deconstructive, pour into China, leaving the architects, owners, and the first party alike in complete confusion. The architecture design in present China can be deconstructive or modern; it can also be super-realistic or eclectic and so on and so forth. The reason for such a situation lies in the fact that Chinese architecture did not experience the Gothic, Renaissance, Baroque, Rococo or new style movements. Historical faultage brought up a contention of a hundred schools of thoughts, and at the same time caused the lack of value evaluation system in our age. The prosperity of architecture design coexists with the blind spot of evaluation and cognitive errors.

For things of the same style, a glance will be enough for us to tell the good from the bad, and to judge whether it has been give painstaking thought, whether it is refined, or well-considered in terms of proportion. However, things of different styles can not be judged by the same standard, because they are not the same kind and accordingly cannot be compared with each other. It is hard to say which style of construction is good or bad; we can only decide which style would best fit campus construction. The president of each university holds in his hands a piece of land, just holding like a piece of cloth, wondering whether to make it into a Sun Yat Sen's uniform or a business suit, jeans and camisoles, or cheongsams. But the architecture theory circle fond of classification with isms would stubbornly grade jeans and cheongsams as good or bad, rejecting the practical measures with academic high-heartedness, and besieging and intercepting emotional catharsis with rational viewpoints.

In my opinion£¬the history of western architecture developed through history while swinging back and forth between technique and romance, sense and sensibility. The recent oscillation from sense to sensibility occurred at the time when people began to have a critical reconsideration of the one-sidedness and tastelessness of modernism. Under the leadership of Venturi (1925—), the architects of "conservative anti-modernism", proposed that "historicism" and "folk art" are the two paddles pushing forward the ship of modern architecture. They advocated that the architectural creation should "stretch out one hand to the ancient times and stretch the other hand to the public". In the middle and the last part of the twentieth century, postmodernism rose to set off another campaign against modernist architecture through the employment of historical fragments in construction. It is reasonable that historicism, as a part of postmodernism, appeared first in the West. From the standpoint of humanism, it questions human nature, values and meanings in the course of modernization and expresses worries in the aspect of culture.

Holding an open conception of history as such£¬we can soberly realize that the emergence and development of any architectural trend is neither the sole product in the field of architecture nor determined by any gifted master. It is comprehensively affected by social, cultural, scientific and idealistic factors. The architectural trend is no more than a spray in the spring tide of history. If we look at any transect of history£¬we can easily find that no architectural form does not embody profound social influence and bear the brand of times.

The history of universities campus construction in Shanghai, and even the history of Chinese modern architecture both have a close connection with the history of western architecture. We can find proofs from historical contradistinction in order to inquire into the mainstream architecture style of Chinese universities at the present stage. Entering the new Millennium£¬the decision makers of campus construction also start to question spontaneously the stiffness, crudeness, stereotyped pattern of campus structures. In addition, they advocate individuation and pay more attention to the cultivation of environment and cultural atmosphere. Their choice of architectural planning betrays their complicated and delicate mood of confusion as they are divided between the modern and the traditional.

The estate giants upholding the banner of historicism, such as Pearl River International, Hopson and Zhejiang Greentown pale the other estate enterprises. The reason, as we can see, lies in their persistent pursuit of the design of historicist style. And their choice of historicism out of numerous architectural styles types is of course based on the strategic design of developers and their own grasp of the market. It reflects more deeply the aesthetic tendency of domestic consumers in present social environment and at the present economic position.

From the view of macro-history, we have made a clear judgment about the direction of the general development of campus building types. Just as Long Yingtai, a scholar from Taiwan, observes, in the process from the traditional to the modern, all efforts are made on approaching "modernization", but instead of putting emphasis on the end of "modernization", the objective of development should be put in the middle of both ends. Just like a piece of "Suzhou embroidery", the two sides, one being science and technology, the other being humanism, should be blended into a harmonious unity.

Ⅲ. The Spirit and Temperament of Campus

People usually give up the pursuit of "loftiness" when they hve acquired "happiness" by means of technique. We require not only "existing" but also "Poetically Residing On". Following blindly the modernity£¬People would neglect the power of emotion, social environment and tradition£¬a power which deserves special attention and a power which make up the essential spirit of a campus.

The theory of modernism is based upon the interpretation of the construction material and culture, with remarkably logical value judgment; while the very core of historicism is the connotation of ideology and culture£¬its value judgment manifesting strong humanism. Before the year 2001£¬the new campus constructions in Shanghai, mostly of modern style£¬are all like stereotyped products on production line. Although the problems in aspects of utility, function and security are solved, their stereotyped forms and colors are really tedious. As far as the campus buildings are concerned, the pursuit of humanist level is above that of the technical level. After we have solved the problems in terms of utility, transport and facilities, we need to consider further how to mould the spirit of campus environment£¬which is often beyond the reach of modernist buildings.

We have incisively seized the spirit and temperament a modern campus needs from the success of constructing East China University of Politics and Law (ECUPL). And with experience gained from it, we are blundering ahead and continuing our steps to design the following items.

As far as the conception itself is concerned, the essence of New Jiangwan Campus of Fudan University is to continue the cultural atmosphere and to shift people's affection towards the old campus to the new campus. Then which form of building can bear the weight of the title "one-century"? When we are rambling on the lawn before Xianghui Hall in Fudan University, near the Great Hall of the University of Shanghai University of Science and Technology or before the Simeng Hall in East China University of Politics and Law, we are extremely amazed by their everlasting glamour. They have been fixed up, but the purpose is to carefully restore them to the original states. On the new Jiangwan campus of Fudan University, the three-piece type proportion, delicate details, elaborate arches and ingenious cloisters of the architectural complex have continued the cultural atmosphere from history to reality, from constructions to surroundings and created a strong rhetoric effect and thus added to the buildings classic charms regardless of the elapsing years. You can praise its profundity or criticize its classicality£»you can attack its aloofness, but you cannot deny its deep cultural accumulation and breath-taking spiritual attachment.

The new library of the South Campus of Shanghai Institute of Technology on the Caobao Road was designed to emphasize the harmony between the new construction and its surroundings and between tradition and history. When a newcomer asks where the new library is, it manifests the highest quality of city design and architectural design, as the new forms a harmonious whole with the old. The new historical viewpoint of university campus architecture should be best embodied by "the successful combination of old and new".

Goethe says, "Constructions are concrete music". We can easily get moved by the magnificence of historicist buildings and are unconsciously brought into a placid and gorgeous state. Construction itself indicates unparallel rhythm. And contrast between softness and heat embodies profound connotations. What is the most important is that the distillation of a historicist construction lies in the expression of human mind or something in people's inner heart rather than the simple utterance. Walking on the old campus of East China University of Politics and Law, you may find yourself become calm and quiet and recall with emotion, reassured by sight of the light building stones and redbrick walls in the shadow of trees.

What do spirit and temperament mean to a campus? As a kind of semiotic system, the total environment of campus carries various meanings, spreads abundant historical cultural connotations, represents the pedagogical tradition and cultural accumulation and is engraved with the development track of a specific historical stage. A construction is the integration of many factors——not only practical, technical and economical elements but also psychological, emotional, aesthetic and psychic elements. It is neither simply an objective event nor subjective consciousness, but is the blending and union of subject and object. Why do we immediately get excited when we think of the constructions in Yale

University and Cambridge University? We marvel at not only the architectural "beauty" but also the gradually perfected cultural connotation and historical marking reflected through history, culture and atmosphere on the campus. A campus with a sense of history can tell people its past glories from which anyone can imagine or experience the complicated emotion of pride, arrogance, loftiness and unremitting pursuit. Looking upon and surveying the architectural style in this way rather than basing upon one's own likes and dislikes might be the due mind-set the decision-makers and architects should hold in the flighty and rash society as ours.

Ⅳ. The Creation and Value of Details

To a building or a monomer, it is details that determine its success or failure. We can find good or not so good, beautiful or not so beautiful products no matter they belong to modernism or historicism. Even though the modernist products are characterized with " less is more"£¬their nodes between glasses and steels and the cut up of dominant and recessive glass curtain walls all manifest their emphasis on details. Modernism without paying attention to details is rude modernism and historicism without paying attention to details is vulgar historicism. The emphasis on details depends on not only architects' professional ethics and quality but also the universities' capital investment and controllability. It is a difficult problem how to build perfect construction with the less money and get more success in cost performance.

The details under discussion are not merely details on the rendering effect. The more important details are those that can be controlled on the execution drawing and can be controlled in the budget. Most decision makers often tell architects that "there is nothing that cannot be done, but there is something you cannot think of". But when they themselves survey the variously colored graphs of effect over and over again, few of them carefully research into the execution drawings (examine the full-page proof of the wall, the division of external wall and the steel structural joint). The rendering effect is brilliant and the appearance of the finished building makes almost no difference, but much is lost. What is missing in the process? Details, details that can be appreciated, touched and perceived and details that are close to human dimensions!

What is the reason£¿In the stage of designing the rendering effect, designers can temporarily slip the leash of funds and blaze new trails audaciously£¬but in the phase of execution drawings, the expensive construction cost of vacuum glass, fluorocarbon steel structures and point supporting glass wall system become the unbearable weight of universities. As a result, they cut out the design of nodes and details. With the same curtain walls and aluminum plates, the constructions built in Germany are romantic, light and pure compared with those ordinary, crude and feculent buildings built in China. The key point is that people have overlooked that a Chinese university cannot bear the cost of expressing details at the present stage. Take the clothes for example, clothes made of crude cloth, with rough workmanship and in strange design are only faulty goods sold at the stalls. If we just pursue the so-called original and unique design under the current economic and technical conditions and the level of construction, we can only build faulty products in the field of construction. With so modest pork barrel on the part of a university and the unilateral construction cost being controlled between RMB2500-RMB3000 per square meter, the pursuit of something new and original can only lead a campus to faulty goods£¡Please take a look at the construction cost of the following several universities:
The classroom building at East China University of Politics and Law£ºRMB2500/m^2
Modern and industrial training centre of Shanghai University of Engineering Science£ºRMB2300 /m^2
The Library of South Campus of Shanghai Institute of Technology£ºRMB3300 /m^2

We are also faced with the problem of leaving a good name of our construction to posterity with RMB 2500/m^2. The construction cost determines the technique and building types employed: getting rid of big-sized glasses and heavy-weight steel structures, avoiding being perplexed by mixed graph effects and even the effect beyond the reach of current technology, making flexible use of bricks and concrete to build the essence of the culture and spirits we need, finding ordinary design techniques with fixed construction cost£¬representing the sense of culture and history and creating the atmosphere of continuing culture.

The excellent design should be consistent with the rendering effect or better than the rendering effect. We can see DHGP's constructions of the campus in historicist style. With only concrete frame and bricks, the walls boast well-arranged unevenness, revealing details inviting meditation, blending elegant historicist symbol and handsome modernist skills together, reflecting clear and individual theme and imaginative architectural space with strong rhythm and distinct form.

The few of DHGP's modernist constructions also attach importance to details with bricks and concrete. For example, the exterior walls are usually built with cheap and elaborate front bricks and dope skillfully, instead of curtain walls and aluminum plates.

The last six years have witnessed DHGP's varied and graceful developments in designing. The south campus of Shanghai Institute of Technology (SIT) is composed of vigorous and firm buildings of the so-called "Chinese indigenous style", a style deeply influenced by the Soviet architecture of East European in the 1950s and 1960s. The newly built campus of East China University of Politics and Law (ECUPL) however is stamped by the Gothic style of Western Europe, towering to the skies. Different from ECUPL, Fudan University uses flat-roofed tower instead, which is the extension of the unsophisticated Greek style, containing everything, crossing national boundaries and pursuing the supreme truth, good and beauty. Nevertheless, Shanghai Fisheries University (SHFU) is just like the Chinese ink and wash embedded in the traditional Chinese culture, elegant and beautiful, brimming with elastic artistic conception. Further more, Shanghai University of Electric Power (SHIEP) is the embodiment of modernist sculpture and depicting the academic subject features through the employment of three rational colors: black, white and gray. These diversified designs not only prove the ability of the architects, but also reveal DHGP's character of innovation. Among all the constructions, the historicism-oriented projects are the brightest pearls illustrating the historicism trilogy in DHGP's development.

The project of East China University of Politics and Law (ECUPL), Shanghai Songjiang University-town, proves a milestone in DHGP's history. At that time, China saw many coarse and crude designs imitating the ancient style, and was even overrun with them. However, the new campus of ECUPL, actually adopting the historicist style that is extricating itself from the pure European classical style while dramatically improving itself according to modern technological level and aesthetic taste, forms a style of neo-classicism or neo-historicism. It absorbs the ratio scale of classical European architecture, assimilates emblems with marked features, and even draws on those unique exaggerating expressive emblems of the Gothic style on the one hand, and on the other hand weds those classical emblems and modern materials and techniques to create a sort of brand new charm, a charm illustrating the traditional handcraft on the basis of modern industry.

Thereafter are the library and building for arts on the south campus of Shanghai Institute of Technology (SIT) on the Caobao Road. Consisting of only two buildings, this project is relatively small compared with DHGP's other projects. Small as it is, it turns out the most classical one among DHGP's early projects. The buildings take off all luxury and exaggeration so that simplicity and modesty are left. The manipulation of the details is artfully artless. This project carries on the spirit of the "Chinese indigenous style" in the 1950s and 1960s. And more than that, stylistically they are indistinguishable from the old buildings, just like a masterpiece of Nature, without any trace of factitiousness. Buildings as such are of top grade.

As for the Jiangwan campus of Fudan University, there is no need to reiterate that it is majestic and marvelous. Nevertheless, I am so depressed to hear the comment that it is of the same style as ECUPL. But in my eyes they cannot be more different. One is full of legislative sternness with Gothic buttress and towers pricking towards the sky, while the other features the magnificence and grandness peculiar to Ancient Greek and Roman style, simple but solid, not to mention the divergence of color. Turning it over in my mind, I finally realize that many people understand "classicism" as "non-modernism". However, they don't know that, just like modernism, classical historicism also has varied expressions of characters and temperament—some are vigorous and magnificent, some unsophisticated and solid, some elegant and beautiful, and still some with profound artistic conception. All these characters and qualities invite people to chew over rather than jump to judgment. By the way, to those fond of classicism, the modernist stuff is but the same.

The above mentioned three projects—the new campus of University of Politics and Law (ECUPL), the Jiangwan campus of Fudan University, and the south campus of Shanghai Institute of Technology (SIT) —are the prelude to the DHGP's trilogy of historicism. In that period, the manipulation of details was getting more mature through disciplines, with the perfect command of classic segments, discourses and emblems. In construction cost control, we became more confident about making excellent work with relatively low construction and fabrication cost. It has been proved that using bricks and concrete to depict details leads to better cost performance.

Still someone will inquire: DHGP should not build another ECUPL or Fudan University for all the achievement in designs. Indeed, I well know that the campus without historical marks is no more than the cultural desert; meanwhile, the historical marks vary from campus to campus. As a historicist, I pondered deeply over after the successful design of ECUPL, Fudan University and Shanghai University of Engineering Science (SUES). Discontented with the success already achieved, I was ready to try something new. I began to be aware that the preceding way is somewhat conservative so that I wanted to carve out a new way, a way not so much loyal to a particular style or restricted by a certain approach.

The chance came from the "Shanghai Conservatory of Music" project, the tender for which was a turning point for DHGP's historicist style. Among several individual buildings could be observed the endeavor to express the classical with glass and steel structure work as well as the early mature classical style. Unfortunately, the further exploration was compelled to a sudden halt since we lost the bid for various reasons. Shanghai Conservatory of Music has opened a door for DHGP and thereafter it entered the second phase of its trilogy. At that moment, the new near-harbor campus of Shanghai Fisheries University (SHFU) and the new campus of Shanghai Institute of Technology (SIT) in Fengxian District almost started construction at the same time.

When designing the individual buildings of Shanghai Fisheries University (SHFU), we apply the simple and unsophisticated architectural language to present the openness and civilization of university buildings£¬with the employment of black, white and gray as the key colors to express the implicit, simple and elegant mood of the campus. The architectural details are harmoniously low keyed, which will become newer when time passes. Accordingly, we rejected such materials as aluminum plate and multicolored brick but constructed with rubble ground floor, stone slab walls, steel gray front bricks, black treadle bars, UPVC frame and glass, etc. instead. Time will prove their long-lasing value.

If Shanghai Fisheries University (SHFU) is a fully-painted Chinese ink and wash with ease and verve, the Fengxian campus of Shanghai Institute of Technology (SIT) should be a rich-colored oil painting. The design of vertical face emphasizes the combination of Chinese tradition and Western cultures, adopting the modern expression of the traditional three-piece type; for the vertical face color we employ warm colors with divergent brightness and layering, pursuing the subtle shades of color for an integrated campus image; and the vertical face details shape a "emblem-system of the new Shanghai Institute of Technology" with unique features, in which there exist the artistic conception of Chinese architecture, the solidness of Western architecture, and some modules of postindustrial age, abandoning the purity flaunted by modernism while pursuing the architectural eclectic aesthetics of neo-historicism.

Shanghai Conservatory of Music, the new near-harbor campus of Shanghai Fisheries University (SHFU) and the Fengxian campus of Shanghai Institute of Technology (SIT), these projects represent the second period of "DHGP's Post Historicism." None of them pursue a specific style any more, but all endeavor to form their own styles, which are unique, divergent in shapes, yet similar in spirit. This poses a tougher challenge for architects.

Our vista of the future is the third period of "DHGP's Post Historicism"—the freer, more lively and tasteful historicism, perhaps being the designs of "Yunchen University(YCU), Shanxi" or "Shanghai University of Political Science and Law (SHUPL)" (with more emphasized ratio of modern elements), which are not included in this book, or some attempts in "Shanghai Conservatory of Music"(with the classical icons of steel and grass, not realized yet because there was no suitable project).

What has been covered in the hasty introduction above, as compared with what DHGP's unremitting efforts through these years, is but a drop in the ocean. Tracing back the history, I am filled with a thousand regrets. The trek and hesitation on the road of architectural creation make me have a profound understanding of "Easier said than done". As far as I know the direction I should stride forward.

Conclusion

The whole architectural design industry of China is roaring so dramatically that the souls of young designers rusted away for seeking quick success and instant benefit in that the industry has moved into products age but still far from works age. When an unprecedented opportunity comes, the industry unfortunately is not thoroughly prepared. Chinese industry of architecture design, still at the stage of obscuration, lacks creativity in every aspect, and the same is true of the theoretical research and architectural criticism in architecture. The Chinese local architects must formulate their own value system and theoretical guidelines for architecture and constantly seeking and exploring burgeoning engineering technologies, material knowledge and designing techniques so as to survive the consistently renewing market environment and create architectural modes which could express people's emotions—architects of our generation should inevitably shoulder that important historical responsibility. It is however, in such a time, tougher a task for all architects to shoulder the historical mission of creating exquisite works. The weak creativity of Chinese architects has led to the commonplace and emptiness of Chinese architecture. "It is indeed a tragedy for present Chinese architects that they have rare opportunities envied by their foreign peers but are at the same time on the horns of an inconceivable dilemma. On the one hand, they take on innumerous projects that architects of all ages can never surpass, and on the other hand, however, they cannot undertake the important task to create some spiritual fortune upon which our descendants will pride," said Professor Wu Jiang from Tongji University.

Let's revert to those problems raised in the introduction. Theoretically speaking, they fall into two categories: one is macro planning, and the other, micro detail manipulation, each respectively representing one of the two values of excellent construction: space creation and details manipulation. Why should the decision-makers feel so hard to choose the style of campus? The difficulty lies in the general lacking of social value system, which also harms the architectural designing. However, we ourselves must enact rules for our soul and set up an inner benchmark for measurement of values.

We should neither confuse ourselves with numerous and complicated problems nor divert our attention into minor details but should seize the essence of campus architecture. My ideal is to "construct the historicist university campus imbued with humanist feelings", the ideal which is being realized in the buildings standing majestically on the solid earth. The "post historicism style" of DHGP—emphasizing the architectural context, respecting tradition, depicting details, and imbedding symbol and metaphor in architecture— indeed fits in with market needs and mass aesthetic standards so that its creative works has been accepted and recognized by the society.

Needless to say, for the faculty and students, the historical imprint of the old campus has developed into a kind of complex within their psychological structure and a kind of prototype embedded deeply in their hearts. The cultural and artistic seed will take root and sprout, if there were suitable environment and warm sunshine. This exactly is the driving power of creation and intrinsic topos for DHGP in the design of university campuses. It is also the source of "DHGP's Post Historicism", which rejects the purity flaunted by modernism and pursues the architectural eclectic aesthetics of the neo-historicism. As a result, contrast and harmony so rendered appear distinctly and the details of approachable scale, full and vivid.

What I have found after years of reflections on architecture design has been well-ecpressed by a scholar from Taiwan, Li Ao, who said, "Nothing could be more boring and worthless than being confused and entangled by modernity and intentionally forgetting the history meanwhile." What we want to convey is not the history of developments in designing, but more profound comprehension. It's true that when history elapses with the passage of time, it changes into a space with sagacity—that is what we should do.

We respect history not because we can hereby meditate on the past, but because historical clues stimulate rational thinking on a higher level. If we place our emotions on the campus, we will surely be moved by those historical clues, through which we could imagine the spirit of the whole lively campus. Just imagine, we will die in a century, but those campus buildings with historical clues will tell our descendants that the benchmarks formulated and details constructed by us have been immortalized. (Picture 16)

16

第一章

华谏建筑完成时

Chapter 1

DHGP Completed

Songjiang Campus, East China University of Politics and Law

华东政法学院松江校区

Shanghai, China, 2003.

华东政法学院创建于1952年，校址设在同年停办的圣约翰大学的旧址。圣约翰大学的校园建筑历经百年沧桑，因其古朴典雅、中西合璧的风格闻名于世，被人们称为“约翰式中国高等学府著名建筑群”。

华东政法学院松江校区是上海松江大学城的二期工程之一，位于松江新城区松江大学园区西北部，占地面积约55万m^2。规划范围北至旗天路，南到张家浜，东临龙源路，沈泾塘将校区分成东、西两块。

华东政法学院松江校区规划不仅运用天人合一、崇尚自然的中国空间理念，同时结合了西方的规划理念和建筑美学观念。运用轴线、对景、空间序列等设计手法将自然要素、水域、建筑、院落这些空间构成要素组织形成有机的、可持续发展的校园空间。

新校区规划建筑设计秉承了老校区的校园风貌特色，从老校区深厚的历史文化底蕴中汲取养分，校园建筑整体布局的构成模式以“组群”为特色。强调组群分明的建筑院落空间，突出新校区与老校区在空间形式上的联系。建筑组群之间以绿地和水体分隔，犹如细胞涵养于组织液之中，将生态型校园的特色发挥到极致。为了隔绝城市道路的噪声，在校园外围城市道路至校园环形主干道之间，沿途均设置15~40m的绿化带，形成围绕校园的林荫绿环。绿环以密植树林为主，将校园的建筑组群包围在绿环之中，同时也强化了校园内部建筑的组群关系。

按照校区的整体规划，校园建筑分为三个层次：第一层次由公共教学楼、图文信息中心、学术及行政中心为主建筑群构成；第二层次由各专业学院楼群构成；第三层次由体育活动场所、学生活动中心、教师活动中心以及食堂等后勤楼群构成。

华东政法学院新校区的建筑风格延续了老校区的建筑文脉，体现“中西合璧、古朴典雅和内敛稳重”的性格特征。采用古典三段式的建筑语汇，材料上采用华政人所熟悉的红砖黛瓦；但又不是对老校区建筑或其他古典建筑照搬照抄或简单模仿，而是努力在汲取建筑精神、延续建筑文脉上下功夫，并在此基础上实现创新。在新校区的建筑风格中，可以十分清晰地看到老校区的一些建筑影子，同时在功能上也满足了现代建筑简约、经济和高效的要求，体现了时代的特征。

01 刺破天空的图文信息中心钟塔，不仅是华东政法学院的标志，也是松江大学园区大学精神的象征。
Soaring into the sky, the belfry of the library represents not only ECUPL but also Songjiang University Town.

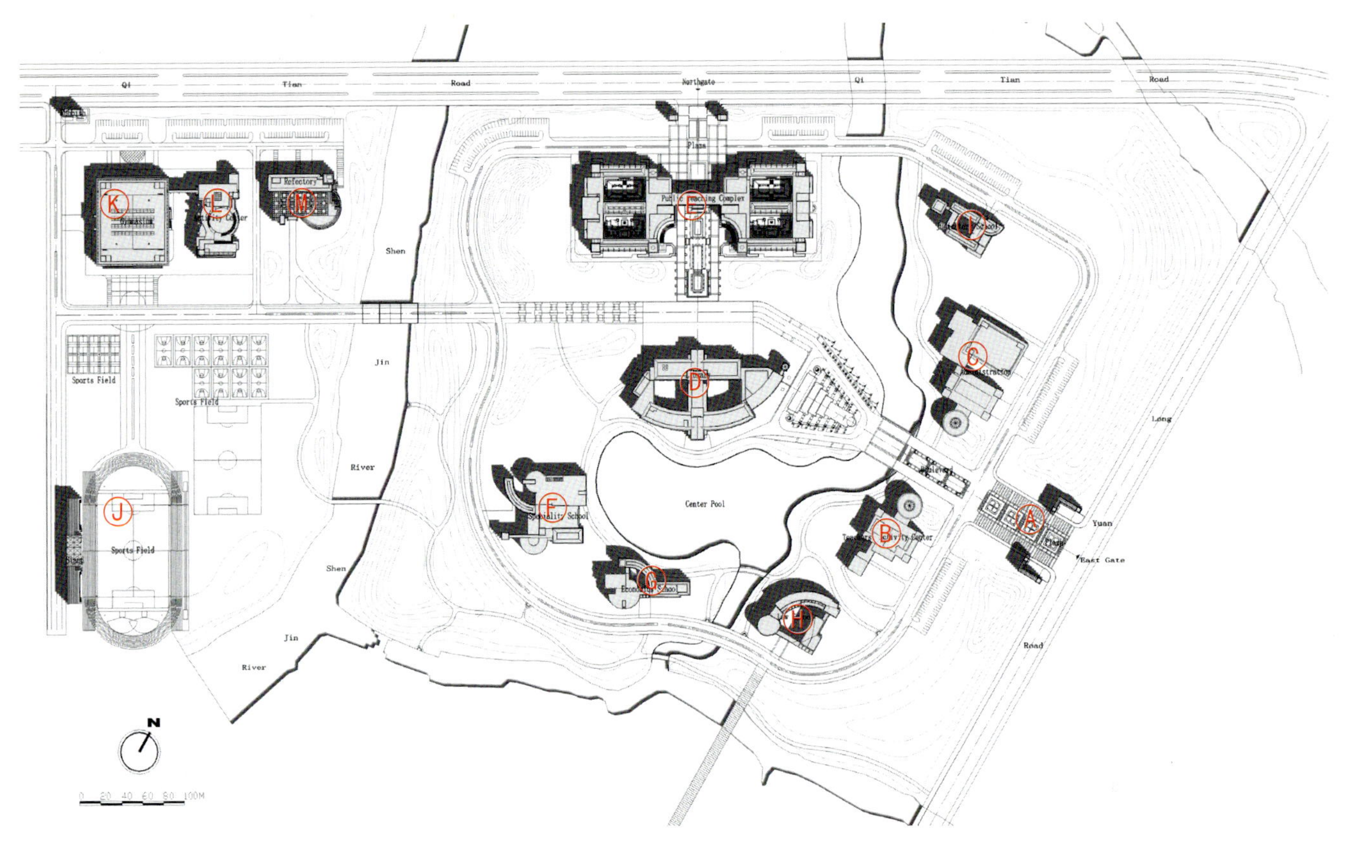

Site plan / 总平面

基本资料		BASIC INFORMATION	
地理位置：	上海市松江区	Location:	Songjiang , Shanghai
用地面积：	550,000m²	Base area:	550,000m²
建筑面积：	122,684m²	Building area:	122,684 m²
占地面积：	44,210m²	Site area:	44,210 m²
容积率：	0.22	Plot ratio:	0.22
绿化率：	54%	Green ratio:	54%
设计时间：	2003年	Time:	2003

East China University of Politics and Law was established in 1952 on the old address of the saint John's university close down in the same year of. The campus building of the Saint John's University experiences successively the vicissitudes of life of a hundred years. Because its style risen in the world is ancient, plain and graceful and it combines the East and the West as a whole, it is been called "John style in the China high senior university buildings" by people.

East China University of Politics is one of the Second items of the University City, located in the northwest of Songjiang University City and it covers area about 550,000 square meters. Planning area is north to QiTian road, south to Zhangjiabang and east to Longyuan road. Shenjing pond divides the campus into east and west piece.

The plan of the campus is not only making use of Chinese space principle that is combining nature and man as a whole and emphasizing nature, but also combined with the programming principle of the west and the construct esthetics idea. Making use of the design skill of stalk line, view and space sequence to organize these main space constitute factors of main natural factor, water, building and courtyard to be an organic campus space which can keeping on developing.

The programming building design of new school inherits special features of the old campus, taking in nutrition from the deep historical culture of the old campus. The whole building of campus takes "sets" as a special feature, emphasizing on a clear building courtyard space and outstanding the contact of the form of the new campus and the old campus. Sets are separated by green area and water. Its relation is similar to the relation of the cell and the cell liquid, making full use of the characteristic of the ecotypic campus. In order to isolating the noise of city road, except the main trunk highway, it will be established 15-40 meters green wreath along with the outer circle city road to form a green circle of shadow surrounding the campus. The circle is formed mainly by forest to surround the buildings of the campus and enhances the relation of the inner buildings of the campus as well.

According to the holistic programming of the campus, the campus building is divided into three levels: The 1st level is composed by public teaching building, information center and the academic and administration center as the main buildings; The second level consists of each professional college building and the third level consists of the athletics activity place, the student activity center, the teacher activity center the dining room and the logistic building.

The building style of the East China University of Politics and Law continued the building style of the old campus, which shows characteristic of the mixture of East and West, the primitive simplicity and elegance. It adopts the building glossaries of three sect pattern. Its material adopts the red bricks and black tiles which are familiar with people in East China University of Politics and Law. But it does not simply copy from the old campus building or other classic buildings. However it makes great efforts to take in the building spirit, to continue the building arteries and veins and to carry out innovation on the basis of it. The building styles of the new campus see some building shadows of the old one clearly. And it also satisfies the request of simple, economy and efficiency in modern buildings, incarnating the characteristic of the times.

A 主校门
B 教师活动中心
C 行政会议中心
D 图文信息中心
E 公共教学楼
F 非法学专业学院
G 国际法、经济法学院
H 法律学院
I 刑事司法学院
J 体育场看台
K 体育馆
L 学生活动中心
M 公共食堂

A Gate
B Teachers' Centre
C Administration & Conference Centre
D Information Centre
E Teaching Complex
F School of non- law profession
G Economic Law School
H Law School
I Criminal Justice School
J Stadium
K Gymnasium
L Students' Centre
M Canteen

明

02 图文信息中心入口，步入知识圣殿的阶梯。
The door of the Information Centre, the Entrance to the Palace of Knowledge.

用地面积:	21,840m²
建筑面积:	24,013m²
占地面积:	6,160m²
容积率:	1.10
建筑密度:	25.7%
绿地率:	46.7%

Base area:	21,840m²
Building area:	24,013m²
Site area:	6,160m²
Plot ratio:	1.10
Building density:	25.7%
Green ratio:	46.7%

图文信息中心

图文信息中心的内部空间从功能上划分为教学、藏书、阅览与办公辅助用房四大部分。平面布局由北面的信息中心与南面的图书馆共同组成，人流路线组织清晰，功能分区明确。图书馆一层设置了目录厅、报告厅、书库及设备用房，二至四层分设各种书库、阅览室及研究室。信息中心一层设有书库及视听演播、编辑用房，二层为行政办公区，三层与四层设有各类计算机教室。信息中心与图书馆之间以4层高的中庭连接，并通过各楼围合出景观内院。西面两层高的校园变电站与图文信息中心以防火墙隔开并将入口分开设置。

大量人员主要由北面主入口的大台阶直接到达二层门厅及中部的中庭。这里是人员汇集、交流、休息的地方。人们在这里可以一边欣赏宏伟壮观的大堂风景，一边交流学术方面的问题。在这里，室内的空间向室外内院延伸，室外景观在室内延续；室内外空间的界线被打破，相互渗透交织，融为一体。东南向的次入口主要为报告厅服务，与读者沙龙成为会议交流共享区。在这里可以进行高档次的学术报告会。

建筑物南侧面向校园中心湖面设置了另一主入口，使图书馆与中心湖岸休闲步行系统结合起来，同时也方便了联系沿湖岸设置的各院系楼，极大地满足了功能上的要求。

03 强调垂直分割的竖向构架与平面方向的横向扩张，在构图上形成鲜明的对比，清晰地书写出华政的建筑情愫。

Emphasizing upright frame in vertical division and transverse expansion in plane direction, displaying the architectural spirit of ECUPL.

Information centre

The inner space of information center is divided from function into four main parts as teaching, storing books, reading and office assistance buildings. The plane arrangement consists of the information center in the north and the library in the south, thus make the person flow route and functional dividing clear. The 1st floor of the library establishes catalogue hall, the report hall, stack and equipment reserving room. The 2nd to the 4th floor establish various stack rooms, reading rooms and research rooms. The information center on the 1st floor establishes stack room and display and edits room for audition. The 2nd floor is the administration area. The 3rd and the 4th floor establish various kinds of computer labs. The information center and library are linked with a medium count about 4-story high. The buildings also make out a view garden through the enclosure of the buildings. In the west, the campus substation about 2-story high is blocked by an anti-fine wall with the information center and with separated entrances.

Most people go to lobby at the 2nd floor and medium court directly through stairs from the northern entrance which is the place for gathering, communication and relax. Through here, people can change issues about academy while enjoying the magnificent view. From here, the inner space extent to the outside garden, the ouside view extent into the inside as well. The borderline of inside and outside is broken and mixed. The minor entrance on the south-east mainly serves the report hall as a sharing area with the reader salon. People can also have upscale academic report meetings here.

To the south of the building is another main entrance which faces the campus lake, thus links the sidewalk system of the center lake with the library. In the meantime, making the contact of college buildings around lake more easier and meeting the demand of function into a great extent.

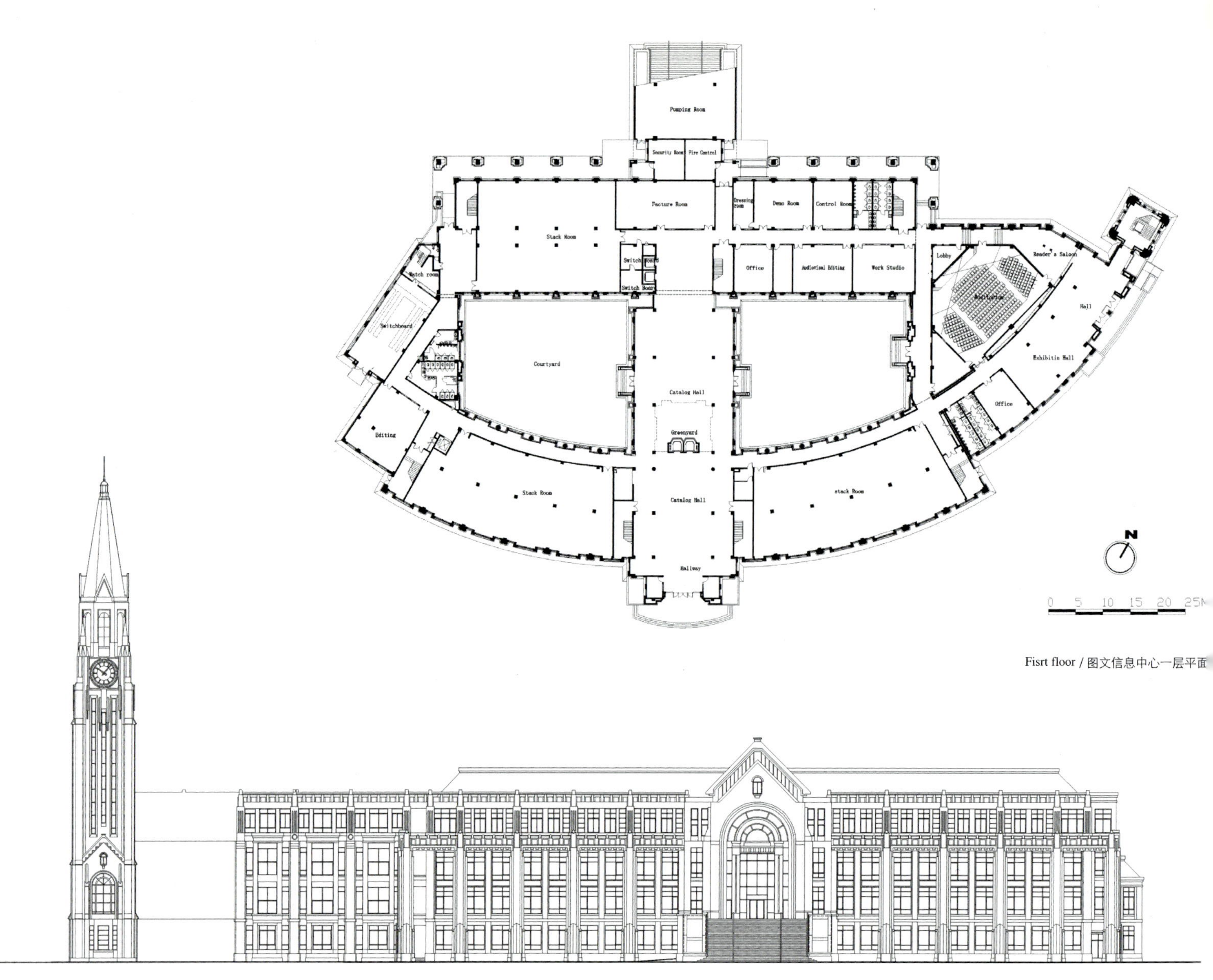

Fisrt floor / 图文信息中心一层平面

North facade / 图文信息中心北立面

04 | 图文信息中心正立面虚实对比，钟塔直冲云霄，大台阶缓步上升，寓意知识的圣殿。

The contrast between reality and vacancy in the façade, the belfry soaring into the sky, the slow-rising wide steps, form an indication of a palace of knowledge.

05 钟塔是整个校园规划的控制点。
The belfry is the reference point of the whole campus.

06 图文信息中心背立面，映衬于碧水蓝天之间，红墙黛瓦愈显古朴典雅。
The rear side of the Information Centre.sets off between clear water and blue sky, with its red tiles and white walls manifesting primitive simplicity and elegance.

07 设计构思草图：不再仅仅是纸面上的精巧构思，已然是现实中矗立的实体。
The exquisite designing concept is no longer a blue print but architectures in reality.

08 从图文信息中心大台阶看教学楼全景，300m 的长卷，诉说着华政百年不朽的故事。

The panorama of the teaching building seen from the wide front steps of the library, the

09 | 课后涌出的成群学生彰显出教学楼古朴沉静外表下脉动的活力。
| After class, groups of students surge out of the teaching building, which denotes its different feature.

公共教学楼

公共教学楼的内部空间从功能上划分为教学、实验、办公与停车辅助用房四大部分，平面以教学用房平行场地进行布置，人流路线组织清晰，功能分区明确。南北主入口设在校园景观主广场侧边，方便人流从室外台阶直接进入门厅。

教学用房共4层，分为东西两部分，东西两楼在一层与四层连通。设计规模为60人教室76个，120人教室8个，150人教室8个，200人教室4个，300人教室2个，听力教室8个。实验部分有100座仲裁法庭和200座模拟法庭各一个。办公用房设于交通联系方便的中间部位。非机动车停车库位于教学用房之下，采用半地下室方式以利采光通风。

The public teaching buildings

The inner space of the public teaching buildings is divided by function into four main parts as teaching, experiment, offices and parking areas. The plane arrangement is focused on classrooms to provide clearly divided functions and proper passage. The north and south main entrances are located by the side of the scenic Main Square to allow convenience of entering the main hall directly.

The four-storied teaching buildings are divided into the east and west parts which are connected on the 1st and the 4th floor. The architecture contains 76 classrooms of 60 seats, 8 classrooms of 120 seats, 4 classrooms of 200 seats, 2 classrooms of 300 seats, 8 listening rooms, one 100-seat arbitration tribunal and one model court of 200 seats. The offices are in the middle part for easy communication. Bikes parking lots are in the basements of the teaching buildings with windows above the ground for lights and ventilation.

用地面积:	45,664m^2
建筑面积:	34,781m^2
占地面积:	8,360m^2
容积率:	0.63
建筑密度:	18.3%
绿地率:	48.4%

Base area:	45,664m^2
Building area:	34,781m^2
Site area:	8,360m^2
Plot ratio:	0.63
Building density:	18.3%
Green ratio:	48.4%

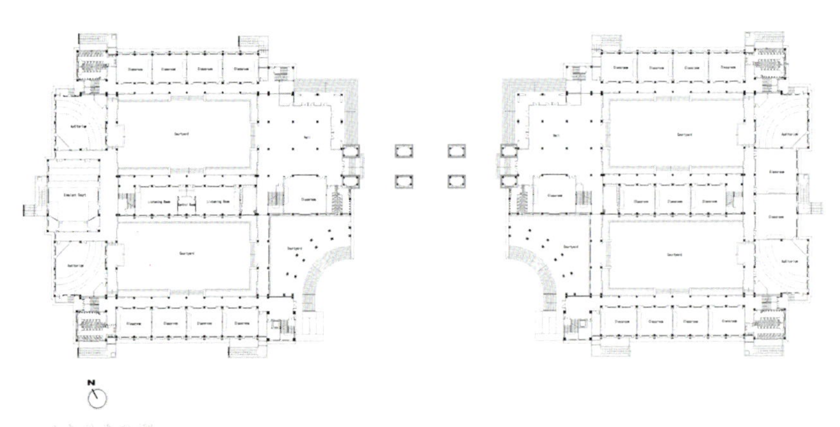

Fisrt floor / 公共教学楼一层平面

10 广场景观细部，场所特质宜人。
Details of the square, pleasant and comfortable.

11 安静亲切的景观水池与古典优雅的建筑语言共同构筑了华政校园迷人的情景，让人们留连其间，感怀沉思。
The quiet scenic pond and the architectures of primitive simplicity and elegance jointly provide a charming picture of the campus in which students and teachers would linger on in deep thoughts.

行政会议中心

行政会议中心从功能上划分为行政办公、报告会议和资料管理三大部分。平面设计上将三大块独立设置，通过走道、门厅和室外灰空间相连，做到既分区明确，又联系方便。

行政办公部分位于建筑南侧，含普通办公室54间，校领导办公室12套，另有中等会议室4间和贵宾接待厅一间。大多数办公室均为南北向，采光通风良好。办公部分以北为报告会议部分，由一个800人报告厅和一个300人小报告厅组成。西北角一层为设备用房，二层为综合档案室。几部分围合出两处全封密的内庭院和一处半封密的庭院，形成丰富的建筑空间。

Administration & Convention Centre

Functionally the Administration & Conference center is divided into three parts: the administration offices, the conference rooms and file rooms. They are arranged independently on plane design and linked by corridors and halls with clear divisions but close connection.

The administration offices are on the south of the building, with 54 clerks' offices, 12 school leaders' offices, 4 medium-sized conference rooms and 1 reception room for honored guests. Most of the offices face the south with excellent lighting and ventilation. To the north of the office part are the conference center composed of a conference hall of 800 seats and a smaller one with 300 seats. In the north-west corner, the equipment room are on the first floor and the comprehensive file rooms are on the second floor. Two whole-hermetic inner yards and a half-hermetic yard are surrounded by these different sections to form an abundant building space.

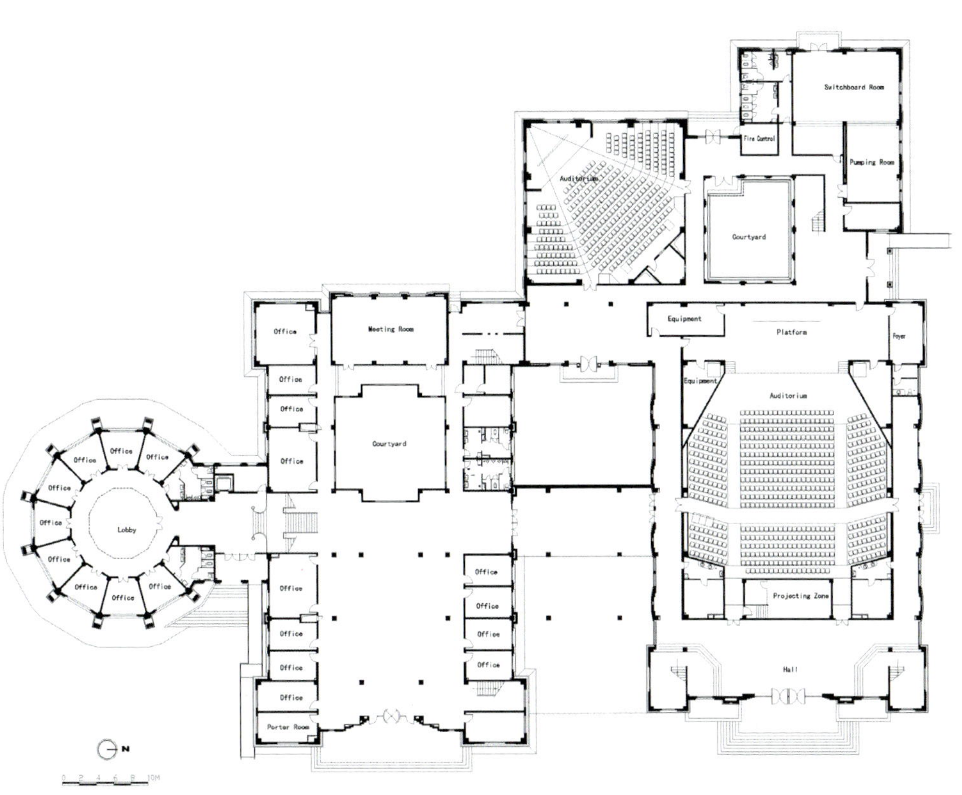

Fisrt floor / 行政会议中心一层平面

12 行政会议中心主入口稳重大方。
The solemn main entrance of the Administration and Conference Centre.

13 行政会议中心的南向透视图。
South view of the Administration and Conference Centre.

14 圆形会议厅的外墙细部，精巧的壁柱使整个建筑散发出典雅的气息。
Details of the external walls of the round conference hall. The exquisite pilasters emit an elegant atmosphere.

用地面积:	16,396m²
建筑面积:	9,740m²
占地面积:	4,187m²
容积率:	0.59
建筑密度:	25.5%
绿化率:	43.3%

Base area:	16,396m²
Building area:	9,740m²
Site area:	4,187m²
Plot ratio:	0.59
Building density:	25.5%
Green ratio:	43.3%

15

教师活动中心

教师活动中心内部空间从功能上划分为客房、餐饮及辅助设施三大部分，平面设计中将三种不同的功能分解成三个独立的体量，三部分通过过道联成一体，形成既分又合的格局。各部分人流组织清晰，互不交叉，功能分区明确。

教师活动中心拥有各种客房60余间，餐厅营业面积近1200m²，同时拥有200m²多功能厅一间，可满足学校交流与接待的需要。

15 教师活动中心的内庭院，圆灯、木凳、卵石，共同构筑了一个沉思的空间。
In the courtyard of the Teachers' Centre, round lamps, wooden chairs and cobbles construct a space for mutation.

16 教师活动中心次入口，隐约可见背后图文信息中心的钟塔。
From the sub-entrance of the Teachers' Centre,Part of the belfry in the back can be observed.

Fisrt floor / 教师活动中心一层平面

Teachers' centre

The internal space of the Teachers' Center is divided by function into three parts: bedrooms, the dining hall and the assistant facilities. In the plane design the three different functions are independently arranged but the three parts become a whole with the connection of corridors while each allows for an easy passage and clearly distinguished functions.

The Teacher's Center owns more than 60 various bedrooms; a dining hall of 200 square meters and a multifunction hall of 200 square meters to meet the needs of communications and reception.

用地面积:	12,224m²	Base area:	12,224m²
建筑面积:	6,072m²	Building area:	6,072m²
占地面积:	3,800m²	Site area:	3,800m²
容积率:	0.50	Plot ratio:	0.50
建筑密度:	31.1%	Building density:	31.1%
绿地率:	52.3%	Green ratio:	52.3%

当代中国一种折衷主义建筑的现实策略

——华东政法松江新校区设计后记

引子

华谏国际（DHGP）在经手华东政法学院松江校区项目的同时也经历了一次自我实现与升华。华东政法学院松江校区建成后，多方面的溢美之辞，反倒触发了我冷静的思考，设计中的苦辣辛酸透射出华谏国际（DHGP）设计团队的思辩和感悟。我挑选了自认为具有挑战性的角度来解读华东政法学院松江校区项目（以下简称“华政项目”），试图以当前整个社会文化环境为背景、用历史主义的眼光，看待华谏国际（DHGP）在华政项目上采取的现实的折衷的建筑策略，体会华谏国际（DHGP）在当代中国建筑界所做的实践和努力。

我得承认这是我长时间以来的思考和自我辩论，而事实上这能不能成其为一个问题，抑或这是不是问题的重点，也许仔细体味生活，热爱建筑的人会有真正的共鸣。

正文

华东政法学院老校园是圣约翰大学于1879年开始建设的。“一八七九年，美国圣公会在沪西梵王渡购地九十亩，建筑两层楼之中国式斋舍一所，住宅四所，这便是最早的圣约翰书院，一九零六年发展为著名的圣约翰大学。”（引文出处）美国圣公会在进入中国时采用的是中国样式的建筑形式，其后的校园建筑带有明显的中式元素。1894~1913年间主要是中西混合的建筑样式，韬奋楼（怀施堂）、格致楼、思颜堂（四十号楼）、思孟堂、大学办公处等，都是在殖民地外廊或古典三段式立面上附加轻飘飘的中式飞檐，很不地道，形态也不是很美，常被形容为“中西合璧”的开端。从1916年建成的罗氏图书馆开始，圣约翰的建筑有了明显的中国传统复兴风格的特征，这之后的建筑比例更适宜，显得更加优美和谐，无愧于“合璧”二字了。

采用中西合璧的建筑形式是西方教会进入中国的一种姿态，一种主动融入当地文化的姿态，这样易于被中国老百姓接受，被地方文化所认同。选择异己的文化来表达一种姿态，这种文化策略显然影响到圣约翰大学校园的建筑策略。高高翘起的嫩戗屋角在西方人眼里就是最具中国本土风格的建筑特征，这样的形态代表着西方人眼中的东方和东方建筑。如果以地道的传统中式眼光来看，这无疑是可笑的。但今天，它作为见证近代上海文化发展的历史文物被仔细地保存着，这种可笑丝毫没有妨碍它成为重要的历史遗存，或者减少它的历史价值。因为它的价值不在于建筑艺术上的特殊地位，而是在于它代表的文化意义。

有趣的是100多年后的今天，华东政法学院异地新建时也选择了用异己的文化来表达一种姿态，选择了与本土文化的背离。同样，文化上的策略引导了建筑的策略，直接地说这是老校区历史文脉的延续，又是松江新城英式风格的规定，然而，追求“新”、追求“不同”于其他校园，才是其选择相异文化来进行自我表达的深刻原因。我们可以发现“创新”一词自始至终出现在设计文本之中。而用历史形式是怎么创新的呢？华谏国际（DHGP）的建筑师们在新校园建筑中灵活地运用和转换西方建筑元素，通过手法主义实现历史形式下这一“创新”，这与当年西方人在圣约翰运用嫩戗屋角也是一致的。

在建筑理论学术界对松江大学城一直保持着审慎态度的同时，松江大学城如火如荼地建设着，而华谏国际（DHGP）在华政新校园的建设中也声名鹊起，得到了教育界和社会的广泛接受和好评。松江大学城整体乖戾的风格引发的文化争论无疑是深刻的，而华东政法学院松江新校区在这其中却有着自己清晰的姿态。

漫步在松江大学城的华东政法新校园，仿佛见到了一个哥特复兴的校园，隐约暗含着耶鲁、普林斯顿等美国名校的意象，然而它又那样崭新，尺度又如此宏大。这样的体验和场景在

现在的社会环境下具有什么样的文化根源呢？

观察华政松江校区的建筑语汇，有说这是哥特复兴，也有一股脑地将其称作新古典主义。事实上，当前的建筑实践不可能再用西方建筑历史中的特定历史时期的既有概念来称呼，因为我们的时代和环境已经完全不同。但从建筑设计的态度和策略来说，华东政法学院松江校区是带有折衷主义意味的中国当代建筑实践，这和西方19世纪的折衷主义有不少相似之处。

西方折衷主义出现在18、19世纪西方建筑方向的迷失和彷徨当中，建筑师们希望从历史的形式当中寻找到新的方向和道路。这一阶段的历史是纷繁和复杂的，潜藏在折衷和手法之下的是严肃的思考和探索。艺术建筑实践中的折衷主义和历史主义所强调的相对性有着某种显而易见的关系。折衷主义是18世纪早期对于历史兴趣发展的产物。在某种意义上，折衷主义是体验历史的一种现象。在历史主义这种意识的语境中，折衷主义有两种表现形式：不同的形式并列存在，就像人们能够在哥特废墟边上看到古典的神庙；某种风格可能会代表某种主流的道德观念，也可能会和社会改革的观念相联系。18世纪下半叶的法国，社会改革的愿望引发了一场回归严肃古典样式的运动。从列杜（Claude Nicolas Ledoux）的建筑和大卫（Jacques Louis David）的绘画中可以体会到。

与西方社会的这一时代相比，现在的中国建筑界也有同样的彷徨和迷茫。中国20世纪末的开放使国人的眼界得以打开，各种西方建筑风格被同时引入，成为符号并存在于大众的认识当中。风格并置和文化模式化、符号化充分方便了国人的“拿来主义”，也引起了国人对西方文化的兴趣。在18世纪的西方社会，古典传统风格的理解随着对希腊建筑的考古和研究发现而扩张，相伴而生的是对于哥特建筑与东方建筑的兴趣。在这两种情形中，社会文化扩张的动态是相似的，都形成了酝酿折衷主义的土壤。

当前决策者们频繁出国考察得到的西方文化理解依然停留在视觉层面。“一城九镇”、“大学城”等，在决策层面显然高于我们当前的讨论，但其认识层面却低于我们的专业讨论，就是在这样的背景和由此主导的文化下，华政项目建筑自然成为了有折衷主义倾向的权宜的建筑。

在中国当代社会里，无论是公共权利的象征还是私人品味的表达，西方建筑的各种风格似乎都找到了自己的位置。古典主义的形式几乎占据了大部分权力机构，英美老牌大学的学院哥特式的影响下哥特复兴风格在学校建筑中存在空间。18世纪的西方人用折衷主义体验历史，我们则用折衷来体验西方文化。

然而这两者的折衷又具本质不同，西方人的折衷主义的形成，在于对自身过去文化强烈的感情，对某段历史的感知，以及用过去风格来引喻某种诗意的幻想或某种道德观念的能力。18世纪晚期，历史知识被广泛地作为文化模式。折衷主义依靠古典风格的力量而成为观念的象征，这与产生它的文化密切相关。中国社会中出现的折衷主义倾向是建立在西方文化舶来品的基础上的。西方建筑的意义对于西方人和东方人自然是迥然不同的。对于西方人来说柱式代表着人体，代表着秩序，古典主义代表着皇权和集中，哥特复兴代表着民族精神的复兴。这些西方的样式对于非专业的国人来说只是一系列虚词，尊贵、典雅、细腻、丰富、文化。

体育馆

由于文化开放和交流的时间不长，整个社会中针对大众的各种西方建筑风格介绍是简单和粗浅的，这些风

格和形式被剥离了相互关系后被标签化地解说着。在这种情况下，大众作为建筑的最终使用者对建筑风格和元素意义的理解就发生了偏差，形成了一套大众认识。这种在专业认识之外的大众认识显得更加视觉化，简单化。扶壁与圆券、哥特和古典主义这些元素和风格，解读者不同，读出的意义便不同。这种有趣的文化现象，只影响建筑的解读，并不影响建筑的使用。而社会范围的意义偏移使这种大众认识成为不可忽视的文化认识，其现实意义甚至大于了专业认识。

面对这出现在专业与大众之间的语言障碍，学者往往执着于去扭转社会认识，毕竟专业认识才是准确的，而现实的实践者往往主动根据大众认识调整自己的语言体系。在真实的社会中，学者型的这种执着往往只有审美意义，没有现实意义，而华谏国际（DHGP）更多地选择了现实意义。这样的现实调整依在实践上就是折衷主义的倾向和手法。

我们处在一个特定的历史阶段，这种社会认识的初级性正是形成这样的社会文化环境的根源。面对大众认识的与专业认识的偏差，华谏国际（DHGP）选择了现实的态度。华谏国际（DHGP）的建筑风格上的折衷正是这样社会环境下现实的对策。而这种折衷手法引发的细节增加，也给解决当前建筑业制造水平低下问题带来了机会。在华政项目中，通过增加建筑细节来弥补细部精致程度的缺失，成为建造层面上一种低技倾向的明智对策。

用学术界的专家的尺度来解读古典建筑语汇和一般大众肯定不同，这时就出现了社会广泛接受和好评，而专家保持审慎的现象。建筑评论与批评往往无法想像和重现建设过程当中的种种矛盾与困难。

正如著名建筑理论家Alan Colquhoun所言，当你越了解一种历史风格，你越不会去重复它。因此从学者的立场看来，当代的建筑不应再去重复古典的形式，应该采用具有时代精神的现代风格。而事实上，在当前的中国建筑界，现代主义和其他各种历史风格一样，多数情况下也只是一种被并置的风格，各种舶来的新建筑语汇不需真正理解只被作为时尚的外衣。当缺失了真正自身意义的现代主义只空有形式的时候，它变得如此苍白和单调，和其他的历史风格同台PK还不知谁能赢。

华政项目并不是一个单纯的风格讨论，我更希望将它作为当代社会文化背景下一种现实建筑对策的讨论。或许建筑理论界对于古典风格的慎用在专业上具有前瞻性，但在现实条件下华政项目却是完成大众文化整合，全面协调各方面矛盾的成功实践，是一次成功的项目运作。

再审视华政项目建筑设计团队的这一批建筑师，他们现代建筑教育的底子和背景清晰地体现在华政项目的建筑当中。仔细观察这些建筑，它们都具有良好的功能分区，主从空间的布置合理，内外空间渗透流动，流线组织清晰便捷，这些都是现代建筑的原则和手法。有这些因素的存在，还能说这是一群古典复兴的建筑吗？因此，用建筑的外衣判断它的立场未免过于肤浅。

抛开风格，抛开主义，我始终相信用心做建筑是设计的终极境界，华政项目的设计透射出华谏国际（DHGP）精致细腻的设计风格，一个个建筑细节的展示，如数家珍，告知你设计者的感悟和心得。广场的喷水池，图文信息中心的扶壁，甚至每一块地砖，每一片瓦，每一个线脚的对位处理，无不显示出华谏国际（DHGP）对建筑设计至善至美境界的执着追求。这时，你也许真的看到了一种古典的精神。在CAD大行其道，手指在复制粘贴键之间翻飞的今天，这种仔细和执着的确是值得敬佩的。建筑是有神的，密斯说，神就在细节当中。细致和考究的工作，通过细部透出来的就是建筑的神。

建造过程是充满了变化波折的，只有亲身参加过建造的人，只有亲历过矛盾的人才能理

解这建造的传奇故事。建筑师和业主的互动至关重要，双方认识的一次次沟通与调整，终于在理解上达成了一致，建筑师用建筑的语言表达了业主的理想，协调了大众的口味，这种转译的成功显然是需要对受众心态和大众认识的充分把握。公共教学楼在这样艰难的沟通中首先落成，由此赢得了业主的认同与信任，对于年轻的建筑师团队来说是至关重要的，之后的建设就在这一份信任中顺利进行下去了。公共教学楼的塔楼经过多次反复和推敲，与业主的沟通和协调就是在专业认识和大众口味之间寻找一个理解的平衡点，寻找正确表达意义的最合适的建筑语言。这一平衡点一旦形成，便深刻地影响着之后设计的每一步。

虽然，多年以后华谏国际（DHGP）的这一实践并不会成为我们这个时代的代表，但从中完整反映出了这个时代社会文化需求和环境的一个侧面，这又使它在当代具有了典型意义。这使我们意识到自己，并唤起了历史的批判。风格和形式的选择始终是一个被一直讨论下去的问题。其中引发的自我批判也将促使建筑师反思，在建筑实践中的积极作为是不是应该更有时代的理想，是努力让建筑改变社会，还是等待社会造就建筑。建筑师从设计的角度来看华政项目仍然是存有遗憾的，建筑间应该具有更良好的相互关系，每幢建筑都很丰富必定造成整体过于高亢，缺乏了平淡的陪衬就会削弱了中心的力量。

历史既提供了需要批判的观念，也提供了锻造这些批判所需要的材料，而一个不断使我们意识到自己，并不断地唤起历史批判的建筑，是我们今天的目标。我想，华政新校园带给我们的的思索远比那些缺乏自我意识的现代风格建筑要多。当代中国建筑的前景和出路何在？决不仅在学院当中。华政项目作为华谏国际（DHGP）的一次实践和探索，提供了一种折衷主义建筑的现实策略。诚如莫负春（原华东政法学院副院长）所言，“华政的建设者们也将给中国校园建筑贡献一份新的智慧。”

学生活动中心

New Jiangwan Campus, Fudan University

复旦大学新江湾城校区

Shanghai, China, 2003.

复旦大学新江湾城校区位于上海新江湾城内，东临淞沪路，西靠国江路，南侧近殷行路、北侧近军工路的南侧支路。校区规划建筑面积约30万m²，计划分三期进行建设。

一期工程建设内容包括图文信息中心、行政会议中心、公共教学楼、先进材料国家实验室、学生食堂、后勤服务用房、主校门等七栋建筑单体以及新校区室外总体的建设（含道路广场、湖泊河流、绿化景观、市政配套等）。

校区建筑规划景观设计的整体性思路是本设计的出发点。建筑单体作为校区整体景观构成的元素之一，体察建筑之间关联的综合形态，是本案建筑规划空间设计的精髓。

复旦新江湾城校区的建筑风格的确立，经过了历时半年的仔细考证。从复旦邯郸路老校区的建筑语言中汲取符号是一种手法，但不符合复旦人求新求变的性格特征，走极端的现代主义是又一个选择，但体现不了百年复旦文化气息。

复旦新江湾城校区建筑最终采用了20世纪20年代流行于上海、巴黎的“新建筑”——浪漫主义新古典手法。她体现了华谏国际对高校建筑作为文化历史载体之一所应表现的建筑形态的思索，以及对社会大众心理的分析。是华谏国际在新历史主义创作道路上的又一次大胆而有益的尝试。

01 百年复旦，世界名校，典雅的希腊复兴式建筑，彰显复旦的人文主义精神。
Fudan University has a history of more than 100 years and high reputation, Its elegant buildings of Greek Revival style embody Fudan's humanism spirit.

The new Jiangwan Campus of Fudan University is located in Shanghai New Jiangwan Town, east of Guojiang Road, west of Songhu Road, north of Yinhang Road and South of Jungong Road. The new campus is planned to build within 3 phases, the total building area of which adds up to 300,000m^2.

The first phase of constructions includes a library, an administration and conference center, public teaching buildings, state key lab of advanced materials, a students' canteen, a logistic building, a main entrance and other environmental construction (roads, square, stream, lake, green land, scenery and municipal logistic facilities).

A comprehensive program is the basis for the planning and landscaping design of the new campus The buildings are part of the whole scenery and are well connected, which obviously is the core of the spatial design in this project.

More than half a year is taken to establish the architectural style of the New Jiangwan Campus through careful research and consideration. If a certain symbol from old buildings in Handan campus of Fudan University is adopted, it is not in accordance with Fudaners' character of seeking the new and change. If an extreme modern style is used, it is not consistent with the Fudan cultural atmosphere of 100- year long history.

Finally, romantic neoclassical style is applied, which was popular in Shanghai and Paris in 1920s. It reflects the idea of H&J International thinking on architectural modalities of campus buildings that they should embody as carriers of culture and history, and its analysis on psychology of the public as well. It is also another audacious and worthy attempt on the way of neoclassical design by H&J International.

基本资料

地理位置：	上海市杨浦区新江湾城
用地面积：	907,000m^2
建筑面积：	306,500m^2
其中	
一期：	93,000m^2
行政与会议中心：	27,560m^2
图文信息中心：	19,690m^2
先进材料国家重点实验室：	22,800m^2
综合教学楼：	14,910m^2
后勤服务用房：	2,300m^2
学生食堂：	5,710m^2
二期：	91,000m^2
三期：	122,500m^2
容积率：	0.34
绿化率：	68%
设计时间：	2004-2005年

BASIC INFORMATION

Location:	New Jiangwan Town, Yangpu District, Shanghai
Base area:	907,000m^2
Building area:	306,500m^2
Including	
1st phase:	93,000m^2
Administration and Conference Center:	27,560m^2
Information Center:	19,690m^2
State Key Lab of Advanced Materials:	22,800m^2
Teaching Complex:	14,910m^2
Logistic House:	2,300m^2
Students canteen:	5,710m^2
2nd phase:	91,000m^2
3rd phase:	122,500m^2
Plot ratio:	0.34
Green ratio:	68%
Time:	2004-2005

A 行政与会议中心
B 图文信息中心
C 校门
D 后勤服务用房
E 学生食堂
F 公共教学楼
G 先进材料国家重点实验室

A Administration & Conference Centre
B Information Centre
C Gate
D Logistic House
E Canteen
F Teaching Complex
G State key Laboratory of Advanced Material

Site plan / 总平面

03

02 行政会议中心位于校区的核心位置，弧形的主立面，拥抱圆形的中心广场，营造出开放、包容的场所精神。
With its arc facade, the administration building is located in the centre of campus. It encompasses the central square and constructs an open and tolerant spirit of place.

03 04 精心的设计使行政会议中心从不同角度、方位欣赏都各具风情。
Elaborately designed, the building emerges as various configurations in different views and directions.

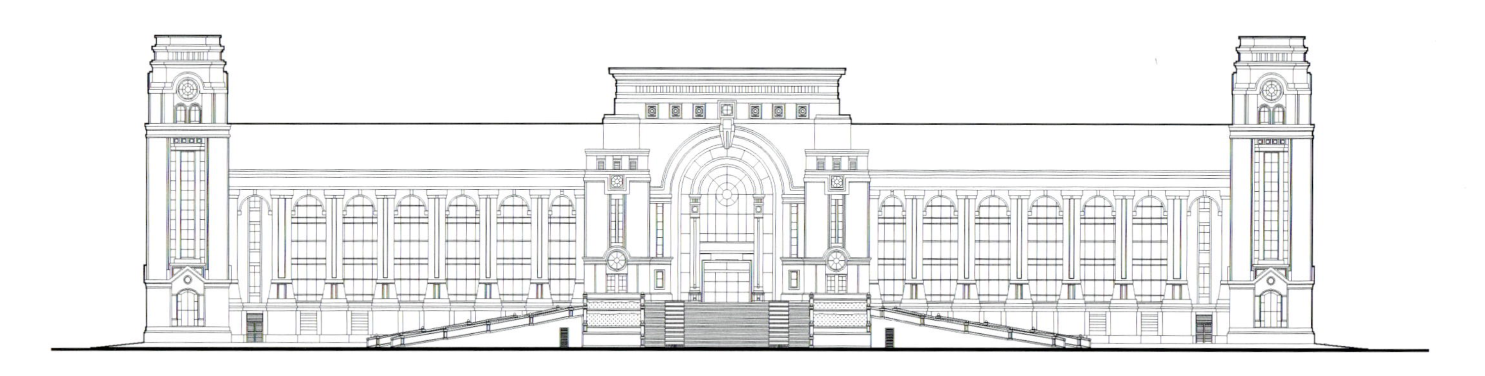

East facade / 行政会议中心东立面

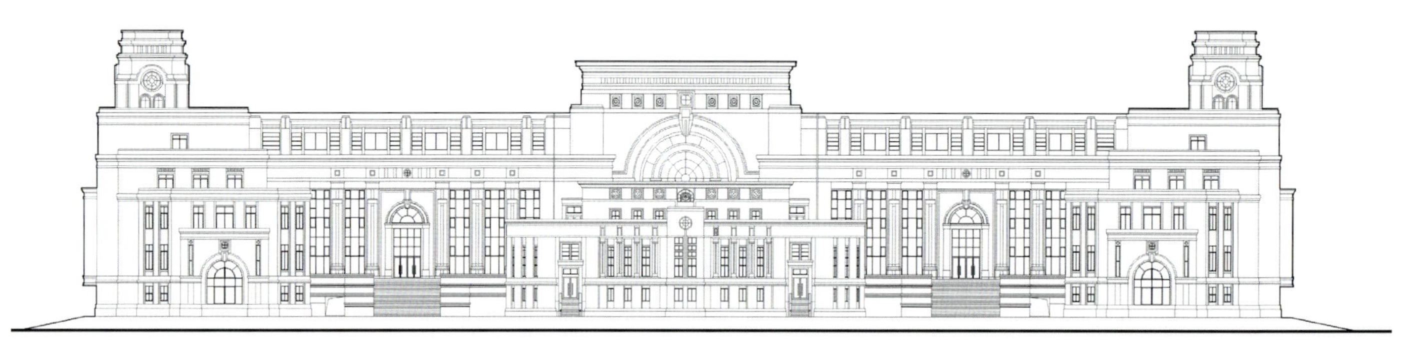

West facade / 行政会议中心西立面

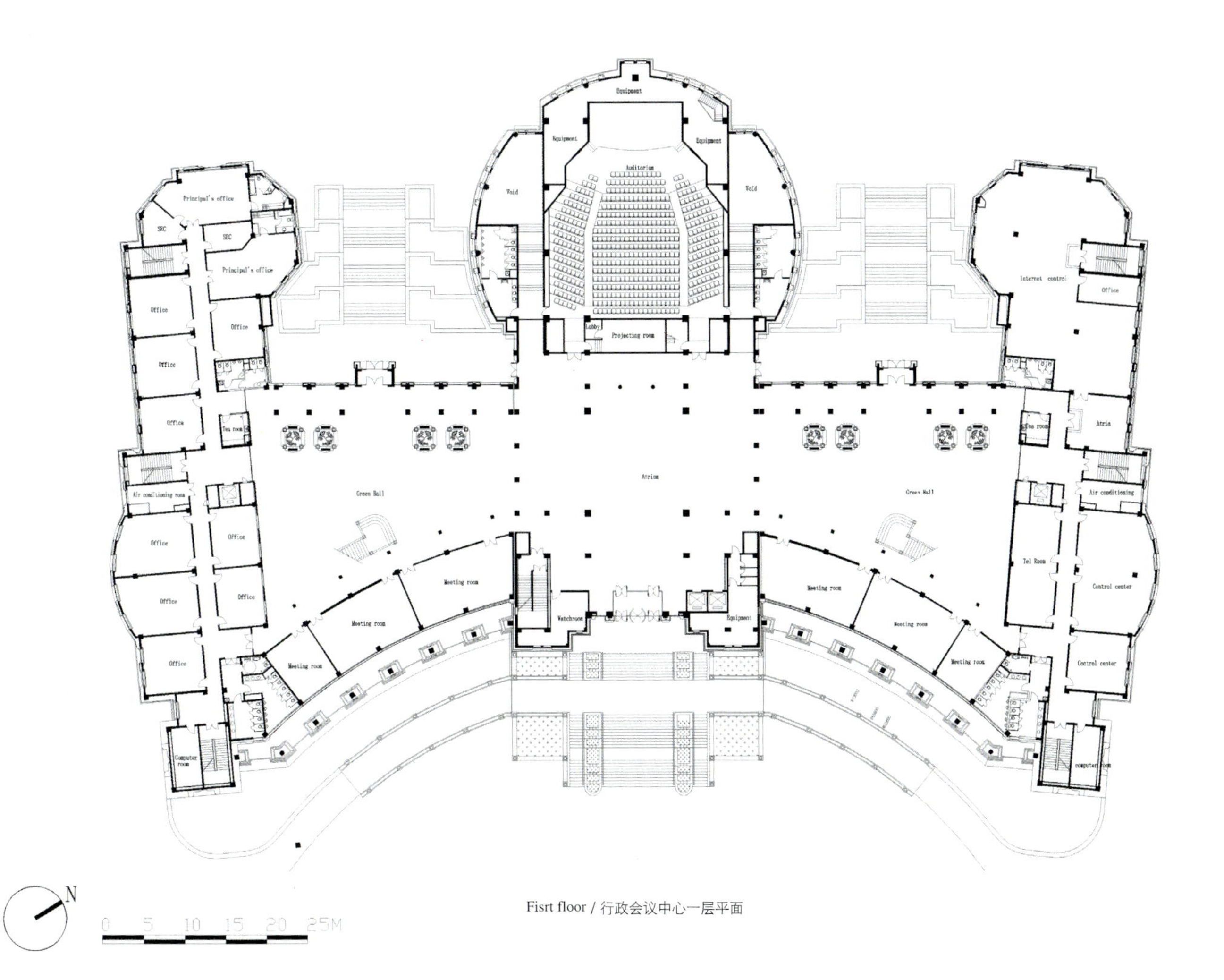

Fisrt floor / 行政会议中心一层平面

05 行政会议中心极富视觉冲击力的环形柱廊，是复旦建筑情节的真实写照。
Ring colonnade has a strong visual impact. It is one of the details in design.

06 柱式、叠涩、圆窗……丰富的细节设计，使行政会议中心的建筑在局部放大后同样经得起推敲。
The pillar, round windows and layers make the design aplenty. It can with stand the observation when it is enlarged.

07 具有高贵的古典气质的吊灯、柱廊，表现出复旦大学室内到室外的整体气质。
The elegant classical pendant lamps and colonnade display the holistic character of Fudan University.

08 光影的变化如同在吟颂一首优美的诗歌，在此你仅仅是忠实的听者。
Transformation of light and shadow is like a poem. Here, you are only a listener.

08

10

09 滨湖的图文信息中心，以扇形的平面很好地适应了基地。
The fan-shaped information centre hear the lake suits the base.

10 从校园核心广场向校园主入口望去，可以感觉到从入口到中心广场微妙的竖向变化。
Seen from the central square to the entrance vertical space is changing slightly.

11 树影背后的图文信息中心。
Information Centre, Standing behind trees and shadows.

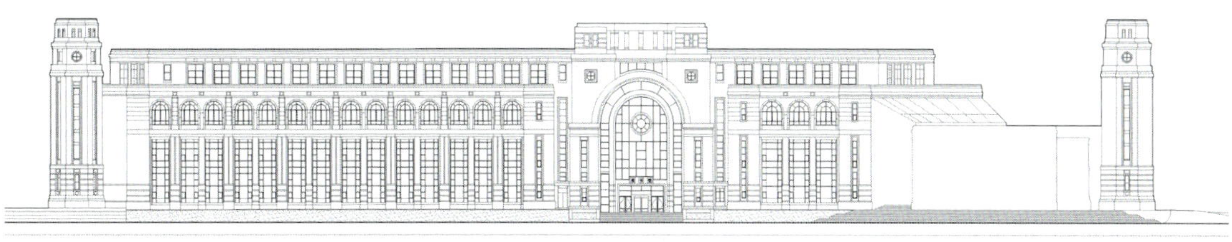

East facade / 图文信息中心东立面

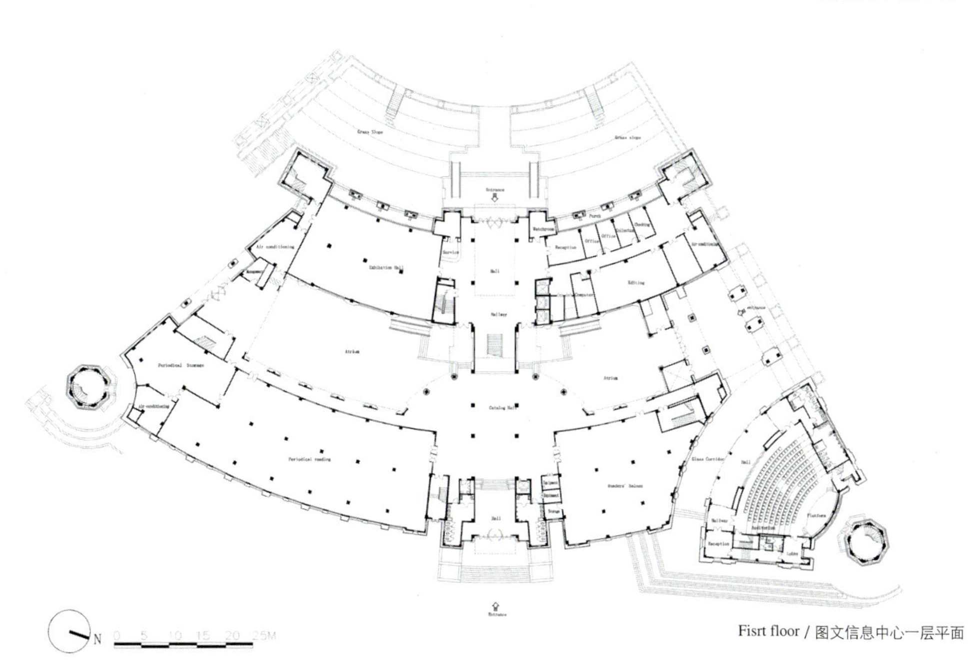

First floor / 图文信息中心一层平面

⑫ 图文信息中心尺度比例精妙的情景化建筑内街。

Interior street: elaborate design in size and proportion.

⑬ 好的建筑会在一天不同的时间，在不同的角度给观者不同的美感。

At any time in one day, a successful building can provide people different aesthetic feelings from different directions.

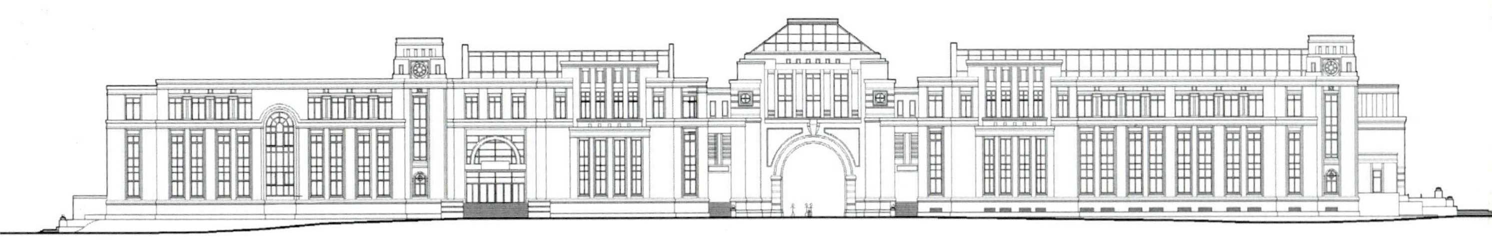

West facade / 教学楼西立面

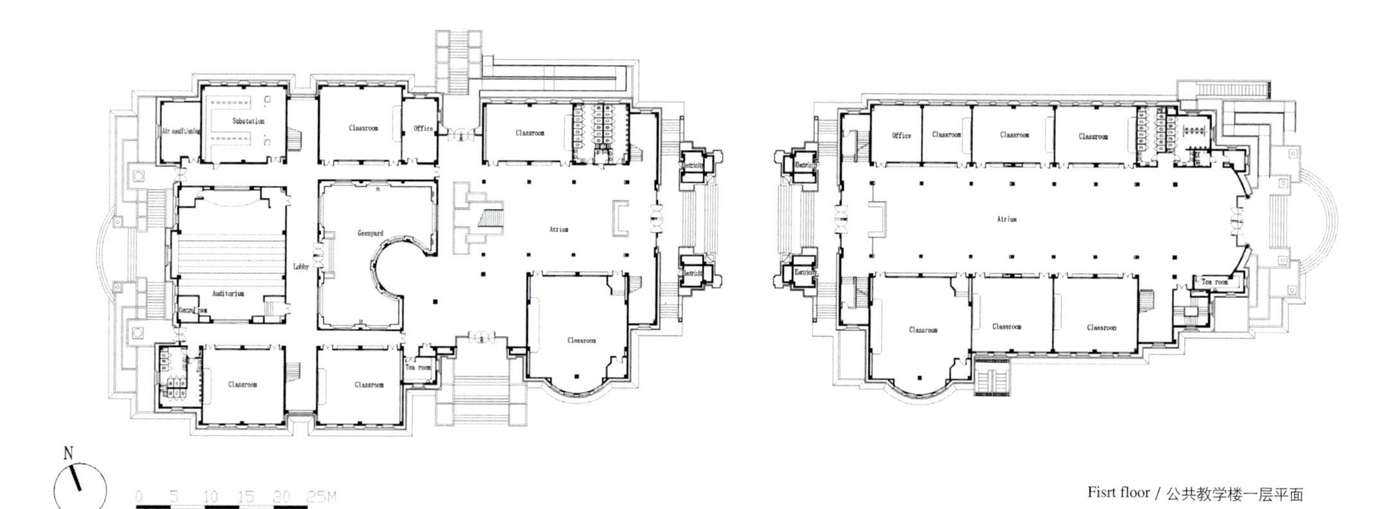

Fisrt floor / 公共教学楼一层平面

⑭ 公共教学楼跨路的门廊，方正敦厚，是校园精神的起止点。
Arch in the entrance hall and its beautiful curve.

⑮ 公共教学楼门廊内的拱顶，曲线优美。
The entrance hall across roads, appears massive.

⑮

16 后勤服务中心内院入口的圆窗演绎出建筑的独特韵味。
The round windows of the entrance accessing to the courtyard of the logistic centre create a unique atmosphere.

17 廊式的主入口门厅空间是整个复旦大学建筑情景剧的序曲。
The corridor of the entrance hall is the prelude of Jiang Wan campus of Fudan University.

18 落日的余辉给食堂抹上了一层金色。
The setting sun lays on gold to the canteen.

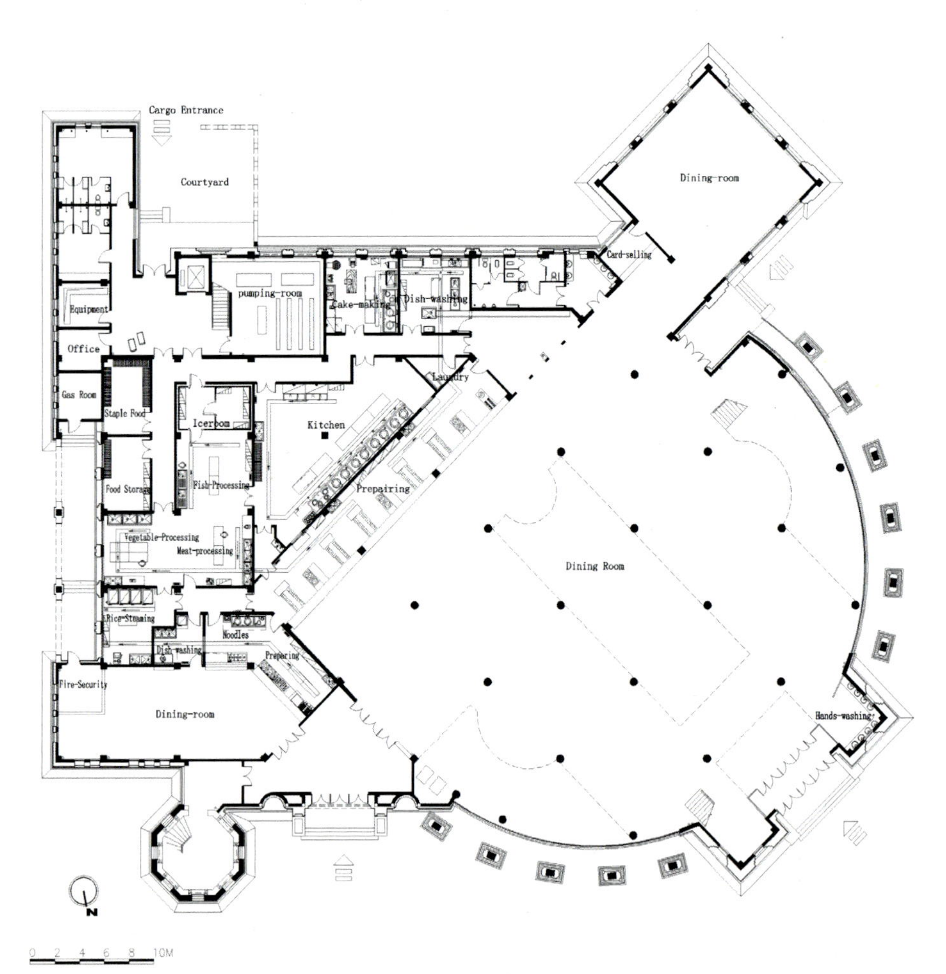

Fisrt floor / 学生食堂一层平面

19 夜幕下的灯光将校门装点得变幻莫测，魅力非凡。
The light set off great charm to the gate at night.

艺术性设计对建筑美学的叠加效应

——从复旦大学江湾新校区看中国大学校园艺术性设计出路

建筑设计的起点是问题的发现，终点是建构直指人心的境界。

过程是什么？为艺术而建筑？抑或为建筑而艺术？

这已经不是建筑技术的范畴，甚至已经超越了建筑艺术本身的界限。

人们使用石头、木材、水泥；人们用它造成住宅、宫殿；这就是营建，创造性在积极活动着。

但，突然间，你打动了我的心，你对我行善，我高兴了，我说：这真美。这就是建筑。艺术就在这里。

——勒•柯布西耶《走向新建筑》（陈志华译）

艺术可能在建筑本身，也可能在建筑之外，有时候人们不置可否，在复旦大学这个特殊的设计个案中，我们将整体校园艺术性设计理解为艺术行为对建筑的叠加效应。

景观设计与建筑设计一起构筑起的空间形态是一种情思，是心灵的反应，她为智者的思想提供了一块归属的天地，而艺术性设计应成为建筑与景观设计之中的点睛之笔，神来之采，是诠释校园历史与人文的落脚点。场所精神的塑造是学校这一类景观建筑创作的核心问题，也是艺术性设计的核心问题。

人们理解的校园应该宁静隽永，富有人情味，既带有古典的优雅和神秘，也传递着现代气息，让人身处其间，感觉到的是深厚的历史积淀和文化底蕴。

我们建筑景观设计关注到了每个师生内心的感悟和情感的寄托。复旦江湾城校区未使用现代建筑的简约叙事手法，而采用20世纪20年代流行于上海、巴黎的“新建筑”——浪漫主义新古典手法。而艺术性设计行为对建筑行为的叠加所表现的是淳朴希腊风格的延伸，承载着海纳百川的包容，跨越民族界限，追求至真至善至美，是对建筑学、社会学、行为学的全面诠释；预言着百年、五百年，上千年的永恒和人世间的正道沧桑。

艺术性设计行为是对建筑语言、建筑性格、建筑空间的再理解……

与环境共生

在艺术性设计的过程中，环境并不是一个中性或消极的因素，而是一种能给艺术家以灵感和创造性的力量，必须倾听环境的声音，并把它看成是作品的有机部分，让作品从具体的环境中生长出来。

建筑与景观共同塑造的空间是真实的领域，当一些精心考虑的物体和线条被引入的时候，就有了尺度和意义，这是公共艺术创造空间的原因。从作品与环境的角度看，作品不仅应与环境和谐统一，而且必须成为环境不可分割的一部分，即艺术性设计的作品对于环境应具有具体性、就地性和不可移位性。

人文情怀

对于校园建筑而言，人文层次是高于技术层次的更高追求，当我们现实中的功能、交通、设备等等问题都被解决之后，我们需要进一步考虑如何塑造校园环境的气质，这需要艺术性设计参与其中。如果说自然环境是实有的、物质的、有形和在场的话，那么人文环境则是虚无的、精神的、无形和空缺的。简言之，人文环境是超越的。由于艺术作品具有超越现实的精神品格，所以，公共艺术与物质环境的和谐，在最终的意义上，应对已有的人文环境给予重构与提升，使其更具有人文的精神与意义。艺术性设计起到的是把人文环境导入自然环境的中介作用，复旦大学新江湾城校区中的每一件公共艺术作品，应以此为原则进行设计。

历史感与永恒性

我们应该清醒地认识到：没有历史印记的校园无异是文化的沙漠。所以我们在艺术性设计上应该有重点地对校园的历史印记作深刻的表现。只有加入有文化的艺术设计才能传承历史，没有历史印记的钢筋混凝土似的校园建筑无疑是文化的沙漠。艺术设计需要注入校园以营造文化品质，这是一个主题设计，需要建筑师与艺术家一同完成。历史记录应该作为一个单独的专题列入考虑的范畴。例如，可在老校区移植一些可供记忆岁月的道具，或中轴大道的铭砖镌刻学生的名字或其他一些历史痕迹，以纪录学校的百年历史和人文印迹。

复旦大学新江湾城校区应该是历久弥新的，正如好酒，愈久愈显露出她的醇香和迷人的魅力，艺术性设计应体现出历史与永恒这一对辩证的主题。

文化精英主义

公共艺术在生活中以其独特的方式呈现出社会与公众的文明程度，同时也是传达一个社会进步思想和观念的艺术性方式。它不是对一个历史形象的简单化记录，而从更深层次的意义上来讲，体现学校性格的一部分。一个地区的文化面貌通常总是以建筑与公共设计作为其硬件标识，无论是古罗马帝国，还是音乐之城维也纳，或是优雅高贵的巴黎，都是建筑与艺术性设计完美结合的典范，她使人们置身于高雅、优美的艺术生存场景之中，使生活在这些城市里的公众沐浴着文明的甘露，同时，以和谐的行为和举止与心境构成文明社会的整体。

从社会学概念来分析，在目前所谓“大学建设运动”中，复旦作为中国一流、国际知名的大学应该以什么样的姿态傲视群雄，独领风骚？从文化精英的概念来概括，复旦人生活和工作的象牙塔应该是一种近人的、宜心的空间；简而言之：不是高档区，而是高尚区；每一块石头，每一个墙面，每一草，每一木都经过精心雕琢，艺术性设计应关注每个师生的内心感悟和情感寄托。

大艺术观

艺术性设计与标识装置设计的难度在于怎样进入，如何与建筑与景观设计合作，一次实质性的深层次合作，而不是简单地做做标志牌子或是雕塑、喷泉。当代的艺术活动不仅把艺术种类之间的界限模糊了，也把艺术的外缘扩大了。请艺术界的朋友们重新思考艺术参与社会活动的使命，设计积极的社会主题意义，并为此拿出最大的诚意和耐心。在大艺术观下，设计的作品关注角度或广或微，表现手法或虚或实，各种艺术形式交融并存，相互穿插对话。

丰富与统一的平衡

标识装置设计作为复旦校园中尺度微小但必不可少的与人体尺度接近的小品类环境，应有统一全局的设计。CI基础设计由复旦大学提供，但如何系统运用CI基础设计，并巧妙与各种装置结合起来，是设计中的重中之重。设计中应注重丰富性与统一律之间的平衡。

在社会生活不断渐进的新世纪，中国当代艺术家迎来了艺术设计的黄金时代，如何以科学精神，严肃、严谨、严格的态度创作艺术精品，以链接城市的文化坐标，延展人类文明，需要我们的坚韧、智慧与共同努力。

雨果曾经说过：“有用的东西是有用的，美的东西是美的，有用并且美的东西则是崇高的。”建筑大概是最能深刻地感受和反映这句话影响的事物了。今天人类的建筑再也不仅是为了使用，更是为了精神层面上的愉悦享受，其源泉就是美的事物和人对美的事物的认识。人，需要在欣赏各种感性美的形式中，获得感觉体验和内心世界充分而高昂的愉快。审美体验是对新校园质量和文化品位的最高检核标准。人只有在艺术美的王国中才是真正自由的。艺术不仅美化建筑，而且美化人类。

Information Centre and Art Building , South Campus of Shanghai Institute of Technology

上海应用技术学院南校区图文信息中心、艺术楼

Shanghai, China, 2003.

上海应用技术学院南校区图文信息中心、艺术楼的建设是典型的历史地段老校区更新案例，是校园功能复兴和结构复兴的重要计划，是应用技术学院在上海高校院校调整后新一轮发展的重要序幕。

图文信息中心采用了纪念性的手法，来传达“知识殿堂”的意象，融入现代建筑语汇，形成了一座与环境共生，具有深厚文化内涵的建筑。书籍依然是图书馆中的基本元素，而图书馆建筑的艺术性和技术性元素在设计中也具有同样重要的地位。

艺术楼则采用相对灵动活泼的手法，打破校园老式建筑肌理结构的束缚，大胆通过建筑线性和界面延伸的具体处理手法，凸现出作为艺术楼这个特殊建筑个体的独特形象，在整体协调中寻求个体变异 。

图文信息中心建筑面积15，800m^2，藏书量60万册；阅览座位总数1，600左右；艺术楼建筑面积约5，000m^2，主要为艺术工作室、阅览室和展示空间。图文信息中心、艺术楼建成后，成为校园的标志性建筑，同时也是校园轴线空间序列的高潮和校园活力的源泉。

01 应用技术学院图文信息中心的入口，用最为端正典雅的方式维系着其作为知识殿堂的高贵与正统。

The entrance of Information Centre remains its dignity and legitimacy as a knowledge palace by the most elegant and decent way.

01

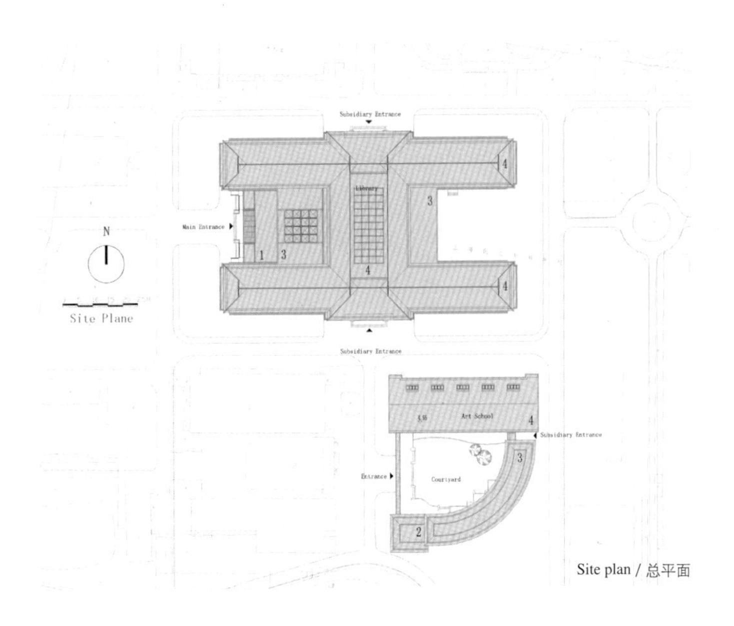

Site plan / 总平面

基本资料		BASIC INFORMATION	
地理位置：	上海市徐汇区	Location:	XuHui District,ShangHai
图文信息中心		**Information Centre**	
用地面积：	8,470 m²	Base area:	8,470 m²
建筑面积：	15,800 m²	Building area:	15,800 m²
占地面积：	4,640 m²	Site area:	4,640 m²
容积率：	1.87	Plot ratio:	1.87
艺术楼		**Art Building**	
用地面积：	5,458 m²	Base area:	5,458 m²
建筑面积：	4,998 m²	Building area:	4,998 m²
占地面积：	1,436 m²	Site area:	1,436 m²
容积率：	0.91	Plot ratio:	0.91
设计时间：	2003年	Time:	2003

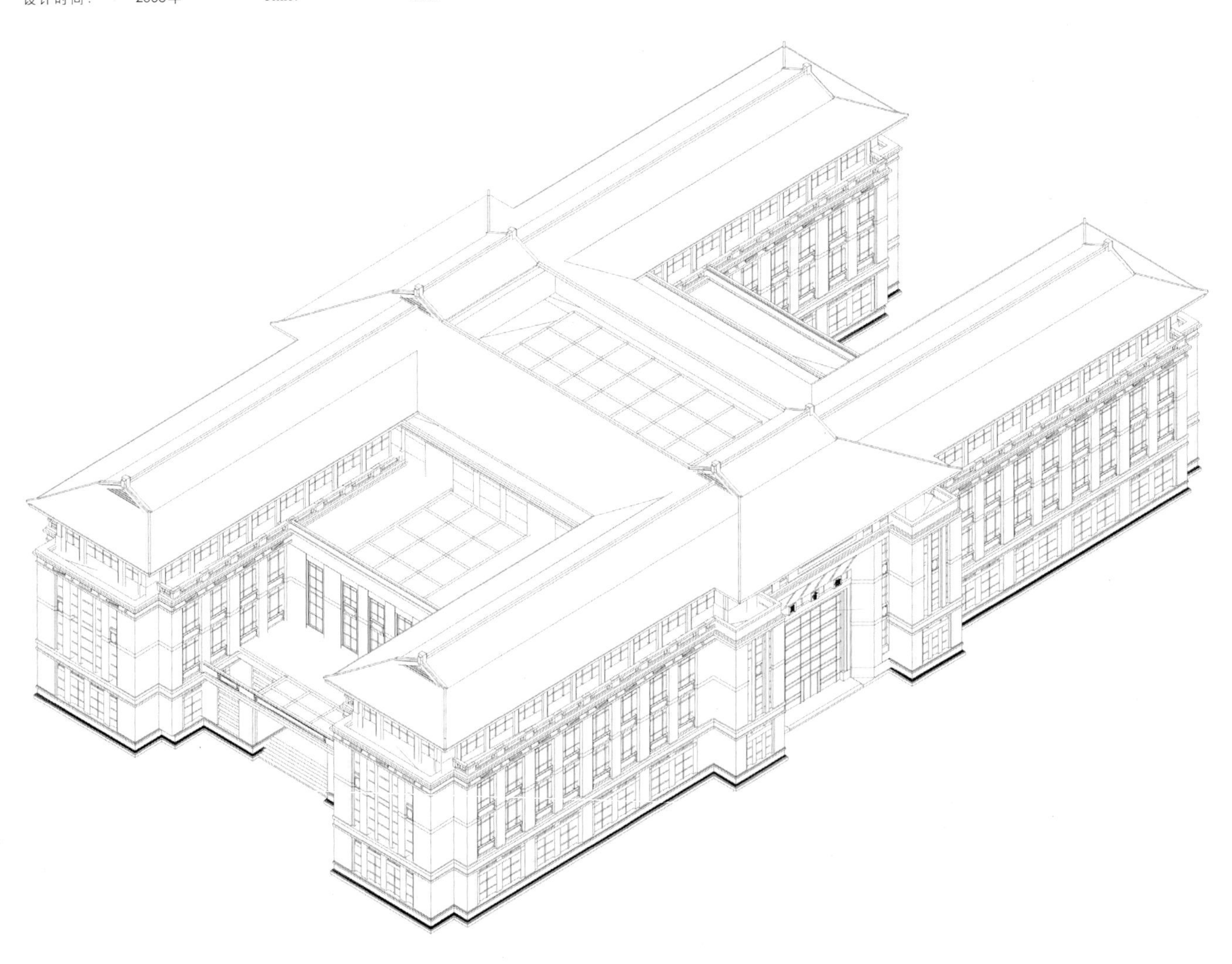

The Information Center and Art Building in the south campus of Shanghai Institute of Technology are not only typical cases in the renewing of old campus in historical district but also representatives of a new stage in the development and reform of universities in Shanghai.

A memorial means is applied in design so as to reserve the features of the "Palace of Knowledge". Various modern architectural elements are added to make the building harmonious with the surroundings outside and the culture inside. Books are the critical element of a library while the features of art and technology of the library building play the same role.

A more flexible method is adopted on the art building. The old style of a traditional school building is casted and replaced by an innovation of stretching the lines and planes of the building. The building's character emerges harmoniously with all other buildings.

The building area of the Information Center is about 15,000m^2. There are more than 600,000 books and about 1,600 reading seats in the library. The Art Building covers an area of 5,000 square meters. When finished, these two buildings will become the symbol of the campus.

02 树影下的图文信息中心。
Information Centre behind the tree.

03 抬头看去，图文信息中心表现出了其神圣的一面。
Looking up,we feel holy from the Information Centre.

04 老校区的建筑掠影。
Briefs of the old campus.

05 应用技术学院图文信息中心、艺术楼设计是一个典型的老校区更新项目。
The project of Information Centre and Art Building is a typical reconstructive project of old campus.

06 午后的阳光洒入图文信息中心，叙述着校园的故事。
The light penetrating the through window of information centre in the afternoon.tells you stories on campus.

07 入口3层高的大厅表达出一种热烈而高雅的气氛。
The three-story hall of the entrance constructs an enthusiastic and elegant atmosphere.

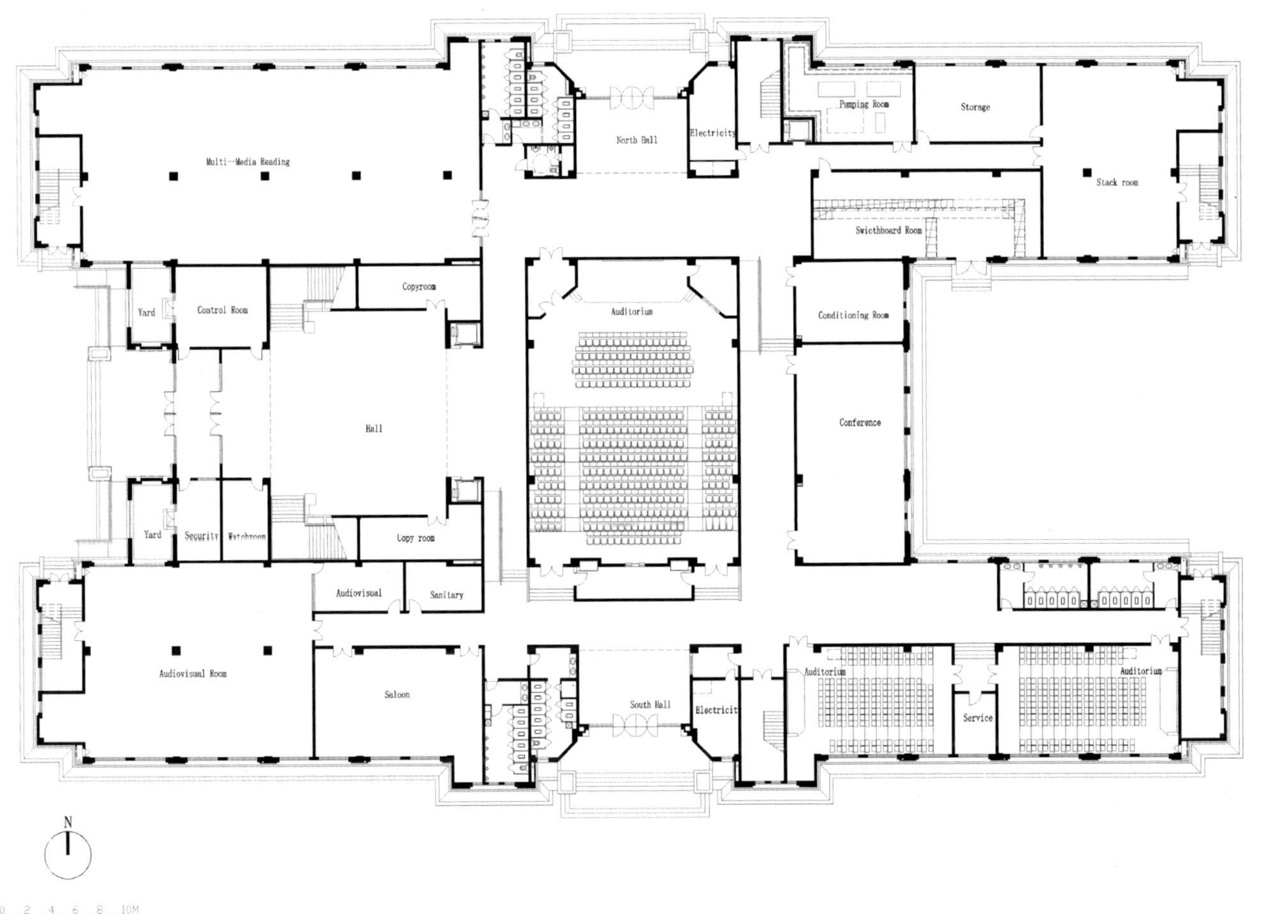

Fisrt floor / 图文信息中心一层平面

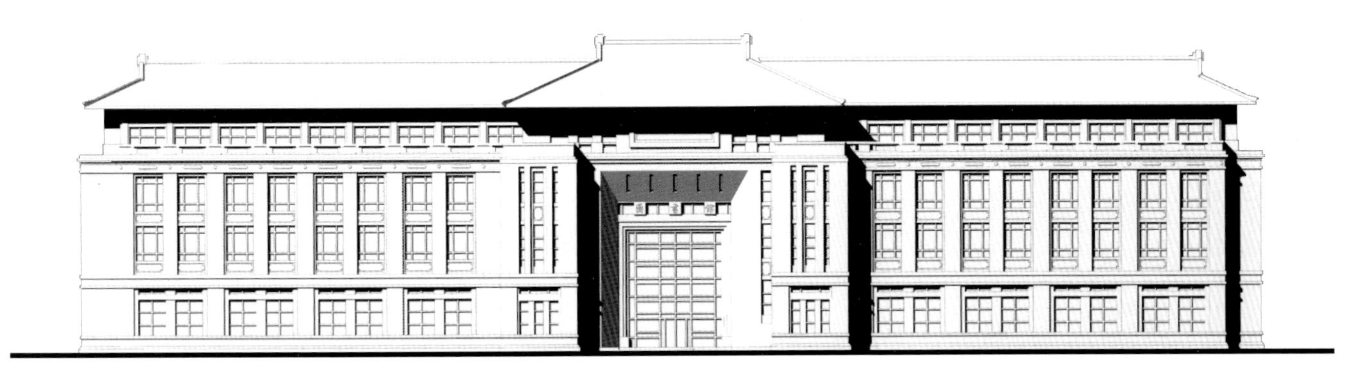

South facade / 图文信息中心南立面

08 看似平淡的单柱廊在镜头下显示出美丽的动感。
The colonnade shows its charm in the shot.

09 单柱边廊并无实际功能，却记录了校园里每一天的日月光华。
The colonnade has no practical functions at all. But it records the change of every minute on campus.

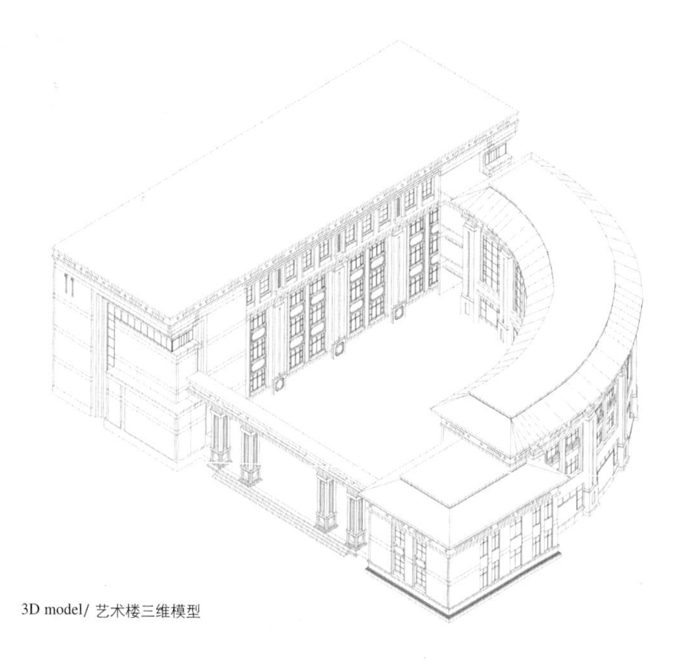

3D model/ 艺术楼三维模型

场所·秩序·细节

上海应用技术学院南校区图书馆、艺术楼设计随想

题记

"在今天，历史对我们已经失去了一部分魅力，在这充满矛盾的世界上，历史已被迫卷入现代生活的潮流。我梦见协和宫空空荡荡，杳无人迹，寂静无声；香榭丽舍大街就像一座安静的游廊。伏尚规划对古老的城市，从圣雪伏到星形广场都不加触动，恢复古代的安宁……人们在这里受教育，生活并憧憬未来：历史不再是一种对生活的威胁，历史已经找到了自己的归宿。"

—曼弗雷多•塔夫里

建构场所

"人必须与其居留处的场所精神和谐一致，才能获得心理上的安定感和满足感。"

—诺伯•舒尔茨

在历史文脉中，创造性的建筑设计可使事物再现岁月流逝所失去的东西，它以一种异化和同化的过程，使我们再一次感受它们的存在，这正是场所精神的本质。

建构我们的场所感要把我们的空间、我们的活动、我们的心理统一起来考虑。场所精神与场所结构是密切相关的。然而作为一种总体气氛，场所精神比空间和特征有着更为广泛和深刻的意义。通过揭示人与环境的总体关系，场所有力地体现出人们居住于世界的存在尺度和意义。在过去，生存意味着与环境建立物质和精神上的良好关系。

我们所指的场所包括五种：

1 物理场所 活动的可能。

当一个项目完成投入使用的头几年，常常会关心我们所设计的空间是否产生了预期的使用效率。失败的作品往往因为其设计师怀着某种自恋情结或者对于某种象征符号的钟爱——包括一些过于庞大的尺度，来概括他假想中的使用者产生的经他设定必须完成的行为。为了避免失败，我们不断地观察，研习一些广受欢迎的场所的传统行为活动模式。

图文信息中心大厅

2 心理场所 活动的可能性的暗示。

文丘里夫人用一句话概括了学生与其他社会人群更具个人意志的特征：当你在一个风景优美的地方为学生放置了一把休闲长椅，他会把它当作床、把脚搁在上面……但是他不会去乖乖坐着——在设计中容易触犯的另一条规律就是：总是错误地估计使用者的心理需求，导致其嵌入不合适的模式中去。

使用者和设计者在空间的使用规则上处于互相暗示的关系，而我们在这场谈判中始终努力地拉近错位的距离，并且改变这个长期以来处于劣势的状况——成为律的契合活动可能的主动

提示者，而非被动的接受设计的失败。

3 视觉场所　视觉规律的契合。

不好的布局总是各有各的混乱，舒畅的视觉场所总是隐藏着某一种规律，而这样的规律是每个人各自对于场所的体验所能够达成的共识。收放、叠景……动用一切能够引起视觉和谐共鸣的手段。

4 文化场所　在前文化经验下对场所的个人解读。

所有的极富冒险精神的远征队在踏上一片新的土地、建造的第一栋楼房、做的第一个规划梦想，永远是原本生活过地区的缩影的翻版实现。场所的熟悉程度会激发人的文化体验，寻根的本能使得所有人在解读事物的时候带有文化印刻，如同法国前卫艺术家眼中的毛泽东在他看来迥然成为一个伟大的浪漫主义导演，代表了每一个表现出的所隶属的那一种文化的意志。

5 社会场所　在前社会经验下对于社会活动可能性的解读从自然人到社会人，人类所处的环境使人们本能的形成一种社会化的行为模式。我们在认同个性解放的同时，社会的共同原型使得我们更加容易分析某一个群体的需求以及他们的愿望。我们的目标就是为他们理想中的生活形态提供场所，使之具体化。

在这里，我们用情感创造诗意空间，体现谦逊的、宁静的姿态。街道的存在为了聚集和交谈，绿地的存在为了健康和运动，学校的存在为了满足学习愿望；而上海应用技术学院南校区图书馆的存在，是为了与整个南校区建筑群落已形成的安静内敛的静谧氛围保持和谐统一。

形成秩序

我们对秩序的追求，体现在图书馆的中轴对称，艺术楼的斜轴对称。

静谧：

在图书馆的设计中极力地在创造一种静谧的感觉，在艺术楼中也同样地弥漫着静谧的歌声。神秘而富有诗意的环境能让人愉悦放松，摆脱烦恼。知识和艺术沉浸在热恋之中，秩序和情感以一种圣洁的安静向敢于问询、倾诉的人传递真理。

平静：

平静是摆脱苦闷与恐惧的良药，无论所设计的建筑是华丽还是简朴，建筑师都有责任在所设计的建筑当中营造这种平静的气氛。无知带给我们与生俱来的恐惧和焦躁，因为这种令人不安的鼓动促成了求知欲。知识就是力量，力量就是从容的来源。营造平静的气氛旨在促进智慧的萌芽，空间秩序传达着哲人思辩拥抱寰宇的影像。

愉悦：

当一个建筑能传达无声的愉悦与平静的时候，这件艺术作品就达到了完美的境界。精神的对话不需要语言来支持，以空间秩序作为中介，以权威式的姿态同时又是平易近人的气质使得每一个知识的朝圣者对其仰望。愉悦的气氛是愉悦学习的开始，建筑秩序传达的完美是对于知识的尊重和心悦诚服。茫然若失到醍醐灌顶的过程，正是所谓的快乐的极致。

关注细节

“细部的设计就像是决定交响乐中的一个音符该如何弹奏，是激昂一点还是温柔一点。总的来说，对于细节的创造，建筑师需要：热情和创造力、自律和毅力。”

“细节是人们能够记住的东西，假如它是独特的、有特色的，人们会记住它。它是一个视觉的东西，最小的细节是在人们的眼睛能及的地方，或脚能踩到的地方。”

现代的建筑师，总是自觉或不自觉地遵循路斯的形式训令“装饰就是罪恶”。过度地追求功能，不仅导致我们对历史文化的摧毁，更让我们失去了集体的记忆：正在消逝的四合院和胡同，都促使我们的设计应回到生活本身。在回眸找寻人文精神的时代，Less is more成为More is more.

细部设计是建筑建造过程中对其形态营造与技术构成的真实体验，是建筑建造与创作表达的完美结合。建筑细部是展现当今建筑技术、反映建筑工艺水平和体现建筑文化特征的有效载体，是评判建筑品质优劣的客观标准之一。细部设计的精度必然决定着一幢建筑的精致程度，是一幢建筑的真实表达，也是建筑师设计思想的结晶。

一幢建筑给人的第一感觉是外部形态及其材料等一些宏观因素，但这些因素都是由许多的细部节点组合而成的，每一个细部设计不仅在结构上确保了建筑的整体性，同样也是建筑师设计理念的载体，是极具象征意义的。欧洲的建筑之所以闻名世界，不仅是因为有先进的技术，更重要的是对建筑细节的高超把握能力。

细部设计还是建筑文化的表达，是建筑风格的一种代表。从建筑的细部可以看出建筑的文化取向，通过这些节点我们可以识别这个建筑的风格，及其要表达的文化特征。

对于细节，中国和西方的理解不一样。在中国，如果描写一个人气度不凡，多赞誉不拘小节；如果描写一个人拘泥于细节，多贬为病态。而在西方，注重细节的人是复杂(sophisticated)的人，它的涵义有赞誉的成分，多是描写这个人非常的考究和不一般的品位。“复杂”这个词在英文里有至少两种意思，一种是我们中文里的涵义(complicated)，另一种涵义就是上面所说的(sophisticated)，是我们中文中不具备的。事实上，细部把建筑区分出来。建筑的细部主要如何使用材料和如何将材料连接起来。如何将它们连接起来是关键的 。

上海应用技术学院南校区的细部营造可谓别具匠心。其校区内老建筑均为中国20世纪五六十年代苏式大屋顶和西洋形制墙身的建筑。在这里，我们通过变异、夸张和重构，以三段式的比例、中式的花窗、屋顶形式的优化和改良，诠释出和谐的建筑图景。

华谏国际（DHGP）在设计中，力图体现古朴无华的情怀，采用了传统的构图手法和人体工程学的“黄金分割”尺度，运用桥廊和中式花窗，强化了建筑的象征性和校园的整体统一感，给人以视觉上的强烈震撼。整个建筑群犹如精心雕琢的艺术品，体现出特有的亲和力和感染力。

当第一次踏入校园的人问及“新建的图书馆在哪里时”，“造新如旧，宛若天成”的设计谋划已在现实中得以印证。

结语

如果说这个项目有何成功之处的话，我们认为那就是对场所、秩序和细节的创造和刻画，从而使建筑散发出迷人的魅力。图书馆是一所大学的神经中枢，储存知识的神庙；艺术楼则好比一块三棱镜，折射多姿多彩的大学生活。当你漫步在校园，你会驻足观赏这美不胜收的景致；在这有限的空间里，你会深深体会到无限的美感。你也会为这巧妙构思所折服。

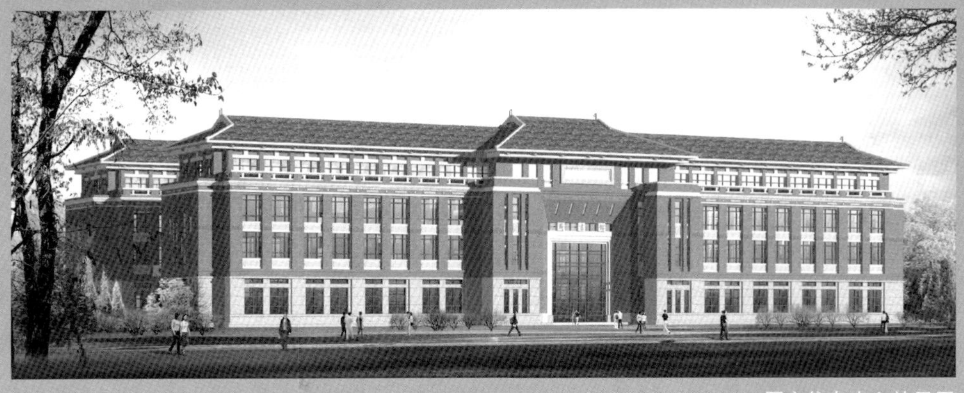

图文信息中心效果图

艺术楼内院

Shanghai University of Engineering Science

上海工程技术大学松江新校区

Shanghai, China, 2003.

上海工程技术大学新校区位于松江新城区松江大学园区西北部，是上海高校结构调整后的特色新兴大学校园。新校区占地面积约78万m^2，规划建设约23.9万m^2的教学、实验、服务用房。规划范围东至龙腾路（与东华大学松江校区相邻），西至龙源路（紧靠华东政法学院松江校区），南到大学园区学生公寓区，北至旗天路。

上海工程技术大学新校区与华东政法学院松江校区相呼应，基地内河道纵横，张家浜、斜路港、荷花湾三条主要河流将基地分割为不规则的四个地块，地域环境特征明确。规划建筑设计合理地利用水系交错的地貌特征，尊重开放性、构成化、多元化的设计原则，整合后的校园水网将校区各功能区块自然分割，并且在水系交汇处放大成湖面，结合校园的景观系统的梳理，形成颇具构成风格的生态化新派大学。

01 体育馆细节实景。
Details of the gymnasium.

The new campus of Shanghai University of engineering science located to the northwest of the university city in the new city area in Songjiang. It is a newly arisen campus after the structure adjusting in Shanghai high schools. The new school area covers about 780,000 square meters area, planed to set up about 239000 square meters in teaching, experiment and service building. The plan is that to the east is Longteng Road(next to the DongHua University), to the west is Longyuan Road(next to the East China University of Politics and Law), to the south is the student apartment area of the University city and to the north is Qitian Road.

The new campus of Shanghai University of Engineering Science and the campus of The East China University of Politics and Law act in cooperation. The base is full of rivers and is divided to four abnormal area by three main rivers which are Zhangjia band, Xielu harbor and Hehua bay. The characteristic of the area environment is certain. The plan of the building design reasonably make use of the geography characteristic that water system is interlocked and respect to open, composite and diversified design principle. Each function of area in the campus is naturally divided by campus water net after integrating, the intersection of water system is enlarged to be a lake and combined with the environment system to form a constitute-style new natural university.

基本资料

地理位置:	上海市松江区
用地面积:	780,000 m²
建筑面积:	230,000m²
其中:	
一期建筑面积:	34,138m²
二期建筑面积:	138,142m²
三期建筑面积:	34,500m²
远期建筑面积:	23,220m²
其中:	
现代工业训练中心:	75,700m²
学生活动中心:	7,700m²
体育馆:	7,980m²
建筑占地面积:	66,450m²
容积率:	0.29
绿化率:	60%
设计时间:	2004年

BASIC INFORMATION

Location:	Songjiang District, shanghai
Base area:	780,000 m²
Building area:	230,000m²
Among them:	
1st phase:	34,138m²
2nd phase:	138,142m²
3rd phase:	34,500m²
Longterm:	23,220m²
Including:	
Modern Industrial Training Center:	75,700m²
Students' Centre:	7,700m²
Gymnasium:	7,980m²
Site area:	66,450m²
Plot ratio:	0.29
Green ratio:	60%
Time:	2004

体育馆:

用地面积:	16,500m²
建筑面积:	7,980m²
占地面积:	5,000m²
容积率:	0.48
建筑密度:	30.3%
绿地率:	35.1%

学生活动中心:

用地面积:	13,032m²
建筑面积:	7,700m²
占地面积:	2,645m²
容积率:	0.59
建筑密度:	20.3%
绿地率:	23.8%

现代工业训练中心:

用地面积:	70,700m²
建筑面积:	75,700m²
占地面积:	20,395m²
容积率:	1.07
建筑密度:	28.8%
绿地率:	35.1%

Gymnasium:

Base area:	16,500m²
Building area:	7,980m²
Site area:	5,000m²
Plot ratio:	0.48
Building density:	30.3%
Green ratio:	35.1%

Studens' Center:

Base area:	13,032m²
Building area:	7,700m²
Site area:	2,645m²
Plot ratio:	0.59
Building density:	20.3%
Green ratio:	23.8%

Modern Industrial Training Center:

Base area:	70,700m²
Building area:	75,700m²
Site area:	20,395m²
Plot ratio:	1.07
Building density:	28.8%
Green ratio:	35.1%

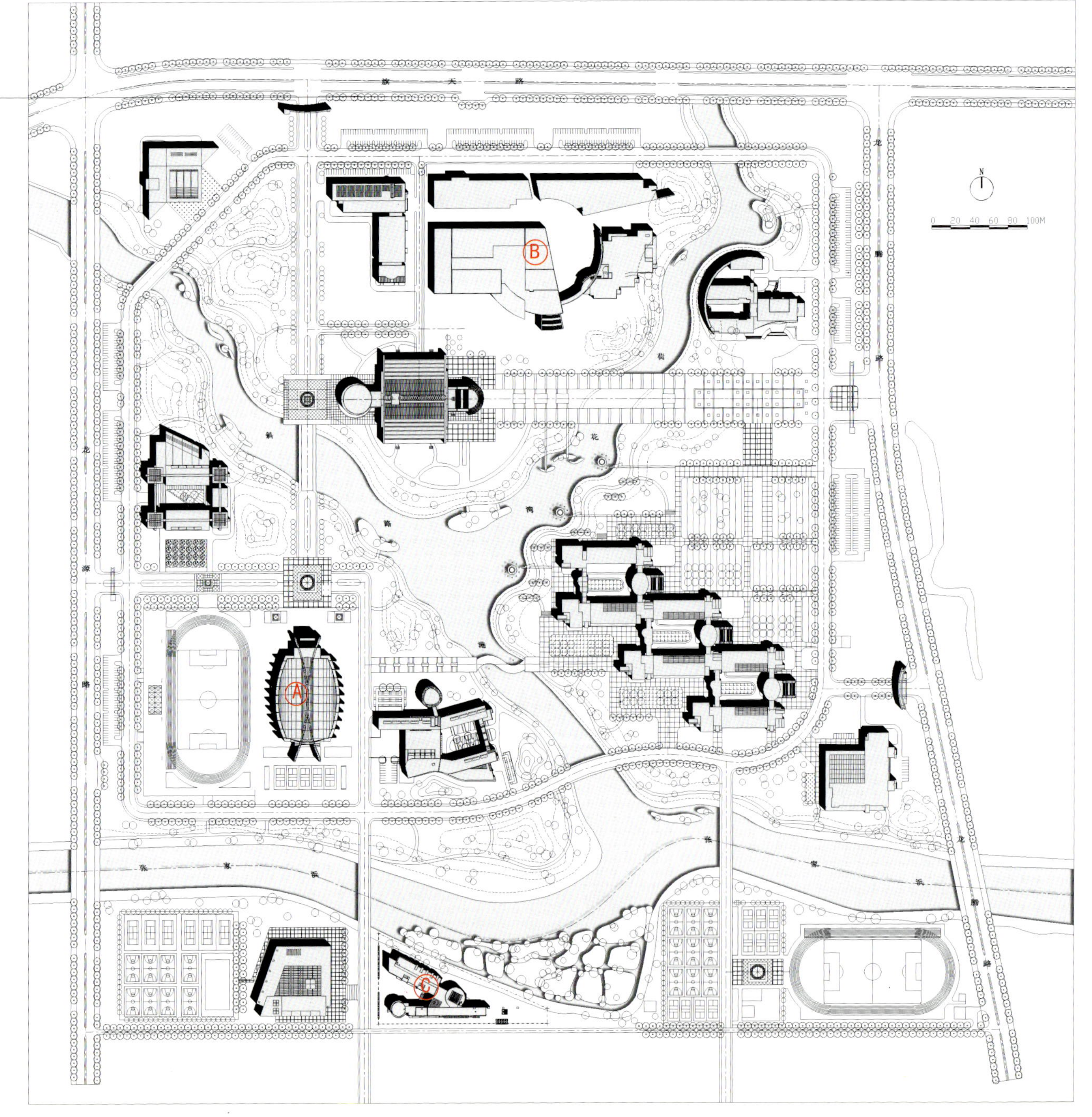

Site plan / 总平面

A 体育馆：

体育馆位于校园南北向次轴的末端，紧靠校园核心湖，和图书馆、教学楼隔湖相望，共同构成了校园的核心建筑群。

体育馆的建筑设计强调了技术美学在建筑中的运用，通过对建筑与景观的整体性设计法则的严格贯彻，合理运用景观建筑学的设计手法，特别突出对体育馆屋顶的营造，表征体育馆强烈的节奏感与韵律感。结构化的理性设计手法和随笔化的感性设计思路的融合是体育馆设计思想的根基，是建筑设计源泉。体育馆外立面的大量钢结构的运用，更加真实地表现出工技大新校园的特质。

B 现代工业训练中心：

现代工业训练中心工程是上海工程技术大学新校区总体规划方案实施中的一组最重要的单体工程项目，也是新校区中最大的建筑群体。作为新校区的标志性建筑群之一，该工程建成以后体现了上海工程技术大学产、学、研相结合的内涵及主要办学特色，并形成以汽车、工程实训为龙头的示范性教学研究区域。

建筑群位于校区东南侧，平面空间布局采用了明显的构成的手法，以一个虚实相间的圆将机械、电子、计算机、汽车等实训楼串联成一体；并以此形成开放性中庭，强化学生间、师生间的交流与对话。设计强调建筑自身的雕塑感，通过形体的组合来表达工业社会中的机械美学原则。理性的设计手法背后是对建筑深层文化意义表达的关注与尝试。建筑材料的运用注重表现材料本身真实特性，石材的质感与肌理更突出建筑的文化特征。不同立面凸凹的韵律随时间的不同而不断变化，时空的转变在此得以被建筑的语汇所记录并传达出来。

C 学生活动中心：

学生活动中心基地位于学校景观河道边，与艺术楼隔岸相望，有良好的自然景观。但用地呈三角形，面积比较紧张；同时紧邻松江大学城学生公寓区出入口，并紧贴桥梁的坡段。用地的情况使设计具有一定的难度，但也为特色的表达提供了可能。

活动中心的设计充分体现了这一特点。设计以巧妙的构思解决了各种矛盾，建筑与基地紧密结合，仿佛从基地中生长出来一样。

建筑平面舒展，使之有最大的沿湖景观面；同时三大建筑体块限定出一个梯形的广场，结合主入口的大台阶，成为学生交流和举行小型演出活动的极佳场所。

建筑空间层次丰富，绿化环境宜人，立面色彩艳丽，造型富有层次和现代感。

A Gymnasium:

The gymnasium is located at the end of the south-north axis of the campus, near the core lake of campus, is separated with the library and the teaching building by a lake and constituting the core buildings of campus.

The design of gymnasium emphasizes on the usage of technique esthetics. By strictly carrying out the principle that buildings and environment designed as a whole, and reasonable using the design skill of view architectonics to stand out the roof design of the gymnasium and to show the rhythm feeling of it. A combination of the reasonable structure design and the sensitive free design is the essence of the gymnasium design and it is the source of building design as well. The usages of large amounts of steel for the out surface of the gymnasium truly show the characteristic of the new campus of Shanghai University of Engineering Science.

B Modern and industrial training centre:

The modern and industrial training center project is a set of the most important single project of the new school area planning of Shanghai University of Engineering Science and it is the biggest building communities in the new school area. As one of the marking buildings of the new school area, the project shows the main school characteristic combining production, studying and research together and starts to take the car and the real engineering training as the demonstration teaching and research district.

The buildings are located at the south-east part of the campus. The organization of the surface space adopts the skill of obvious composing. The machine, electronics, calculator and automobile training buildings and so on are combined as a whole by using a circle combined with truth and imagine, forming an open atrium and strengthening the communication of students and teachers. The design emphasizes on the feeling of the sculpture of the building itself and expresses the mechanical esthetics principle in the industrial society by the combination of form and structure. The rational design is to try and to pay attention to the expression of the deep cultural meaning of buildings. The usage of building materials is focused on the true characteristic of the true nature of them and the texture of the stone material stands out the culture characteristic of buildings. The uneven rhythms of different surfaces change continuously with the change of time so that the change of space-time can be recorded and expressed by buildings.

The new campus of Shanghai University of engineering science located to the northwest of the university town in the new town area in Songjiang. It is a newly arisen campus after the structure adjusting in Shanghai high schools. The new school area covers about 780,000 square meters, planed to set up about 239,000 square meters in teaching, experiment and service building. To the east of the university is Longteng Road(next to the DongHua University), to the west is Longyuan Road(next to the East China University of Politics and Law), to the south is the student apartment area of the University city and to the north is Qitian Road.

The new campus of Shanghai University of Engineering Science and the campus of The East China University of Politics and Law act in cooperation. The base is full of rivers and is divided to four abnormal areas by three main rivers, which are Zhangjia band, Xielu harbor and Hehua bay. The characteristic of the area environment is certain. The plan of the building design reasonably make use of the geography characteristic that water system is interlocked and respect to open, composite and diversified design principle. Each functional area in the campus is naturally divided by campus water net after integrating; the intersection of water system is enlarged to be a lake and combined with the environment to form a new constitute-style natural university.

C Student activity centre:

The activity center is located next to the river for viewing in the campus, separated by the shore with art building and having wonderful scenery. However, the area looks like a triangle, getting close to the entrance of the student apartment area of the University City in Songjiang and sticking to the ascent segment of the bridge tightly. Because of it, the design is difficult, but it is possible to design the campus with special features.

The design of the activity center fully shows this characteristic. The design solves various conflicts with skillful conceiving. Buildings combined with the base so close that they look look like they are born from the base.

The surface of buildings is wide and flat in order to have the biggest viewing surface along the lake. In the meantime, three huge constructions limit a trapezium ground, combining with the big step of the main entrance to become a perfect place for small performance and students' communication.

The arrangement of the Construct space is abundant; the natural environment is beautiful; the surface is colorful and sculpt is full of levels and modern feelings.

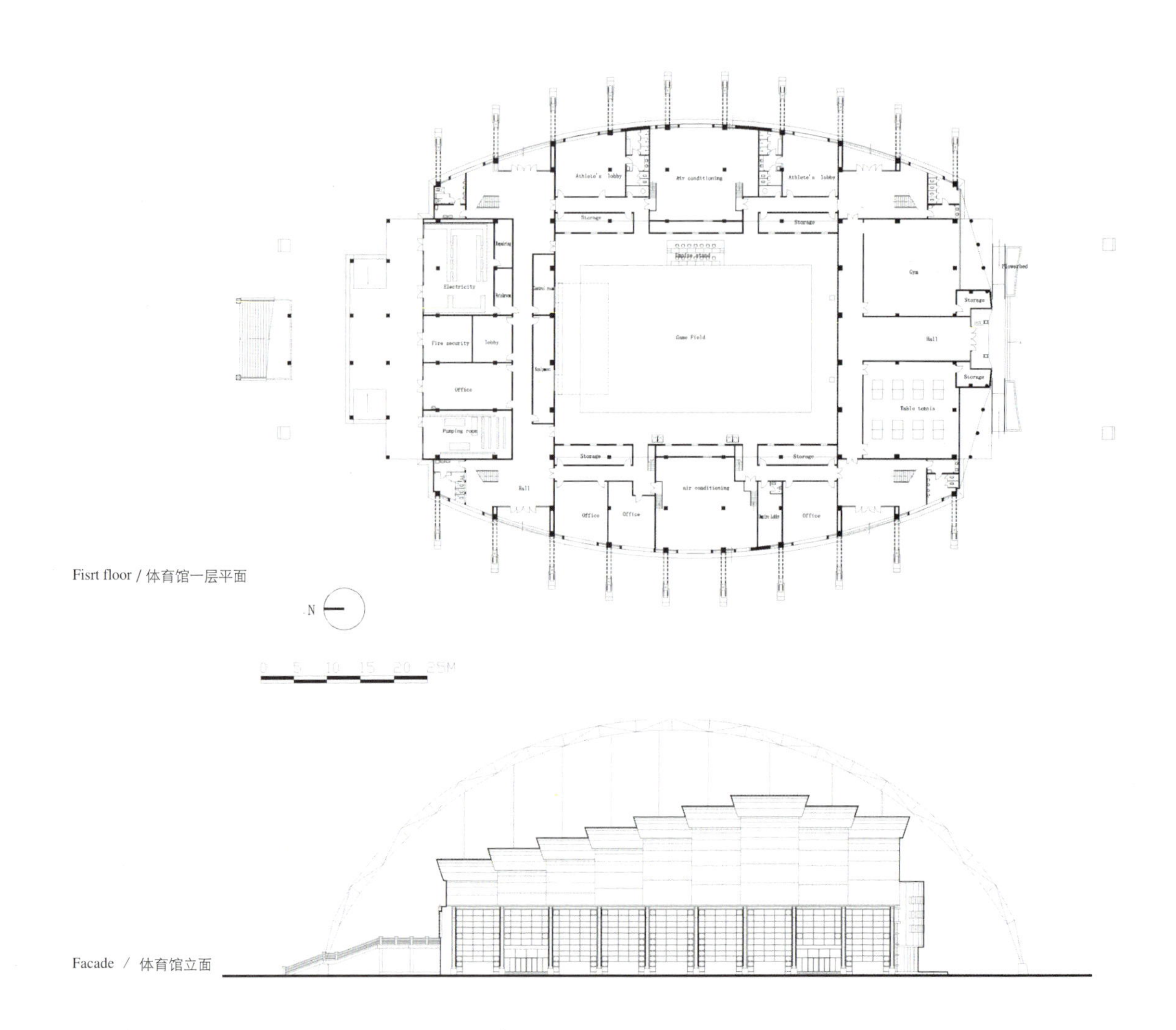

Fisrt floor / 体育馆一层平面

Facade / 体育馆立面

02 体育馆主入口。
The entrance of gymnasium.

03 04 精致的细部构件充分表现了工艺和技术之美。
Exquisite details of the members reflect the beauty of technology and techniques.

03

04

05 融在湖光夜色中的体育馆。
The gymnasium at night.

06 从不同角度看现代工业训练中心的局部。
Modern Industrial Training Centre. Seen from different directions.

07 现代工业训练中心全景效果图。
The panorama of Modern Industrial Training Centre.

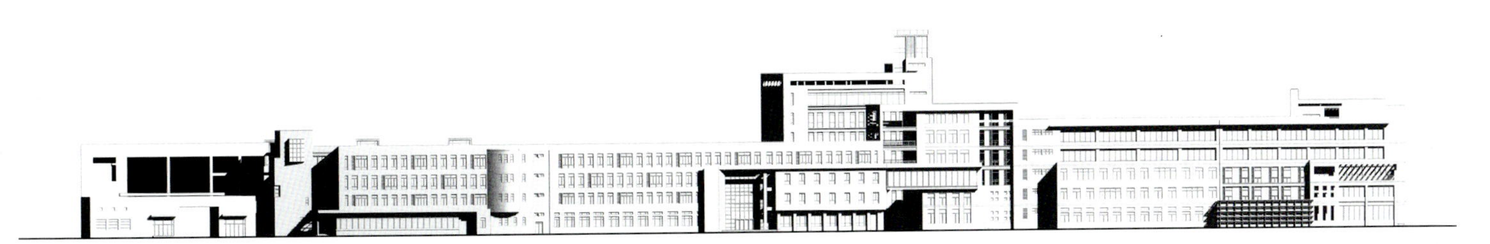

Shade / 阴影律动

08 现代工业训练中心采用现代设计手法，同时局部运用钢结构加以装饰。

We use modern techniques to design the training centre. Part of the facade is decorated with steel structure.

09 现代工业训练中心融入城市设计的手法，营造了一个内向的街景空间。

The design, in harmony with the city plan, creates an interior street in the training centre.

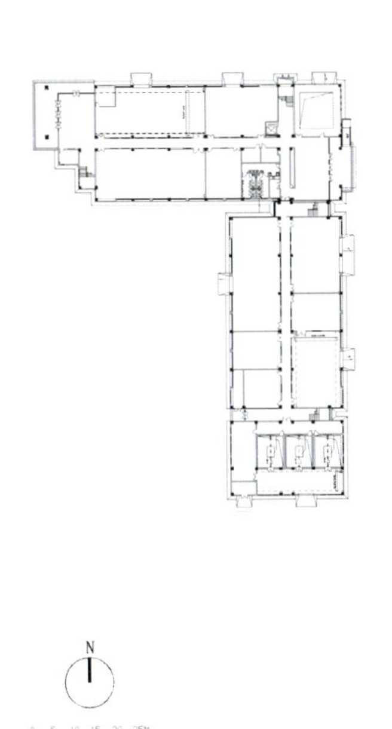

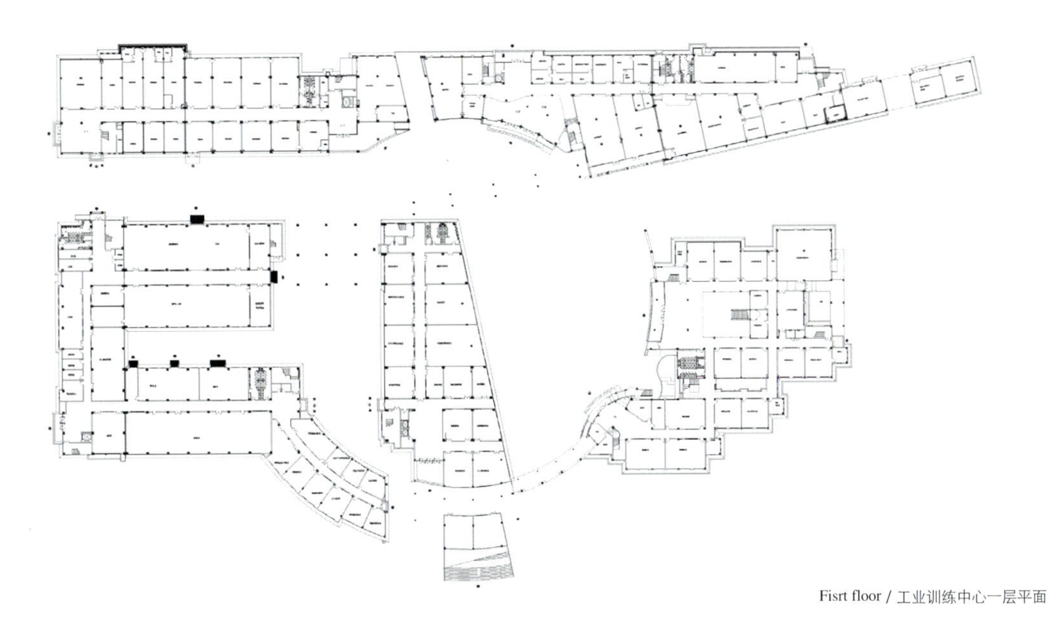

Fisrt floor / 工业训练中心一层平面

⑩ 学生活动中心饱满活泼的色彩表明了建筑的用途。
The bright and vivid colors of the students' centre indicates its purpose.

⑪ 学生活动中心鸟瞰效果图。
Airscape of the student's centre.

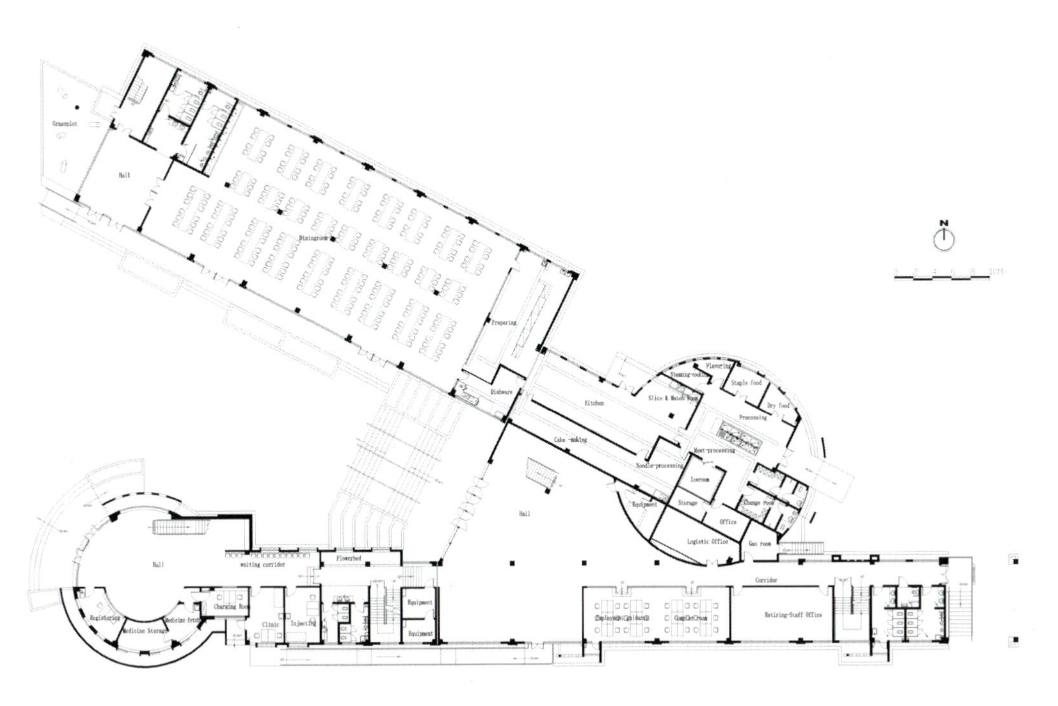

Fisrt floor / 学生活动中心一层平面

⑫

12 挑出部分的水平长窗搭配横向木百叶，这又是一个讲述光的故事的建筑。
The long windows together with the transverse wood louvers in the cantilever part tell another story of light.

13 14 15 16 学生活动中心主入口院落空间的多角度视图。
View from different directions of the student's centre.

高校教学建筑集约化发展趋势

——上海工程技术大学松江新校区现代工业训练中心设计随记

时代的进步对设计与兴建新的高校教学建筑提出了更高的要求，要求决策者与设计者站在时代的最前沿来进行决策与设计。

DHGP 华谏国际的设计师团队在上海工程技术大学松江新校区现代工业训练中心的设计过程中，通过对优秀案例的考察与研究；对新技术与新材料的翻阅与分析；对新的空间形式的研究，发现高校教学建筑有如下的发展趋势：

技术设备主流化

现代教育体系中，不仅有现代的教育思想，而且有先进的技术设施作配合。随着电脑和网络的发展，知识生产和传播的节奏加快，学习方式的丰富和多样化，教学活动和学习渠道得到了极大的扩展。

上海工程技术大学松江新校区现代工业训练中心（后简称现代工业中心）由于功能体系的复杂性，对技术设备的设计及电脑网络在建筑中的运用提出了极高的要求。现代工业中心包括有电子电气学院、化工学院、基础学院、工程实训中心、机械工程学院、材料工程学院等，是一个教学实验综合体，内部含大量不同学科的各类专业实验室，设计师在考虑合适功能布局的基础上，特别在注重对技术设备的工艺设计以及电化网络系统的设计，可以说现代工业中心的各类专业实验设备单元的工艺设计及弱电系统设计已经成为了区划建筑功能的决定要素。

在现代高校的教学、科研等各个领域的工作中，将系统地、全面地使用计算机及网络；学校各项管理工作，也相应以计算机为主要手段，迅速地电脑化，如图书馆的借阅，研究室的出入，甚至是选课，保健都能够通过磁卡进行计算机管理。现代工业中心的设计中，我们也很好地运用了电化网络系统，很好解决了以往传统教学实验中的信息流通不畅、管理不科学的弊病。

电脑和网络在高校中，逐渐成为师生学习、研究工作中的通用工具和不可或缺的重要手段。因此，在高校教学实验建筑设计中，特别是在现代工业中心的设计过程中，设计师们开始充分考虑适应计算机运作的各种技术条件的要求，空间的设计上对设施的定位、布置和安排加于考虑，综合布线、夹层地板、多种照明、系统空调、多方位电源等等。为了全面提高教学、科研的速度、质量和深度，接受系统、遥控系统、特种机器配备的应变系统等设施也成为影响设计的因素。

现代工业中心的设计建造给了我们创造性的启示，对技术设备的考虑和安排应在建筑的设计与建设实施全过程中充分体现，在建筑的艺术设计中不仅需要更多地考虑技术设备科学安排，而更应该把智能化提升为现代教学实验建筑的特殊品格。

室内空间对传统教学单元的突破

现代工业中心的新的教育技术手段的引入和教学方式的改变使教师在授课、解惑的过程中不再仅通过书本、黑板、粉笔的模式，教育电视、投影仪、录音器材、计算机等多媒体及科技设备的引入改变了教学方式，也改变了教学课本，普通教室的形式也逐渐改变。

高校的教学不能仅限于单方面的知识灌输和应试技巧的训练，而应提倡人文教育，知识领域的拓宽，注重人格、能力和个性的培养。教室不再只是学生听讲的课堂，而还应成为他们讨论、交流、实践乃至游戏的场所。个性化的教学，需要高效、灵活而有弹性的教学空间，因此教学空间由固定封闭的小隔间向灵活分隔的大空间转化。具体措施如：柱网较大的框架结构体系的采用，使教室的大小不再受墙体的限制；房间分割更加通用化、模数化，教室、实验室的尺度均较大，可通用、互换；教室的层高相对比较高，以保留出夹层地板和吊顶的高度；吸声材料运用于顶棚及地板对噪声进行合理控制；运用活动家具或屏风等手段在必要的地方提供视觉上的私密性等。

现代工业中心在整体空间设计中提出了“空”的核心概念，在每一个单元功能体中，合理地运用现代结构体系的灵活性特征，为教室空间的多重可变提供了可能，根据教学实验活动发展与进步，教室空间可以扮演的角色是多元化的。设计中特别把大部分需要大型运输车辆的各类大型机械实验室放在建筑物的底层，以便与教学设备的更新。同时功能单元内部，运用轻质隔墙做各类教室空间的分割，以适应学科的不断发展与改革；电子电气学院、化工学院、工程实训中心、机械工程学院等均采用“单位拷贝法则”进行空间设计，为教学改革留足多重可行性。

现代工业中心的空间灵活的教室实际使用上可以适应多种教育方式，如：传统教学、团体教学、混合年级教学等，有利于同专业学生、交叉专业学生、不同专业学生的学业交流；并能够提供更多的新型活动方式和桌椅布置方式，使学生的各种行为能力得以提高，促进同学之间、老师与同学之间的互动性的交流活动。

教育实验建筑在满足基本功能使用要求的前提下，更应该对基本教学单元提出空间复合化的设计理念，打破以往追求功能设计单一化的束缚，突出空间界面的虚化手法。同时综合发挥电化虚拟系统的独特魅力，为教学活动提供三维甚至四维的立体感受。现代工业中心的这种“异质同构”的教学空间设计观念为营造复合化的教学空间作出了有益并且大胆的实际探索。

交往空间的地位

传统的许多高校教学建筑，由于受到功能、经济等各方面的限制，往往设计成中间走廊两边教室的长廊形式。走廊作为汇集所有人流的集散地，仅仅具有交通功能。这种传统的教学空间已经无法满足现代教学的要求——单一地获取知识的学习方式不能培养出全面的人才。

我们可以发现，在信息革命的新时代，大学教学不仅应突出智能生产的重要性，更应加强集体智能的协作与广泛的接触、交流。各种研讨和交流，可以是有组织的、有计划的，也可以是无组织、无计划的，往往无组织的、无计划的交流更经常、更灵活、收益更广泛。因此，能促进师生自发、自由交流的空间环境设计，就显得特别重要了。

随着教育观念的发展，教学建筑的功能也随之产生变化，即在大小不同的功能空间中，有机地穿插交流空间和休息空间。许多新型的高校教学建筑虽仍以教室为核心，但配备有资料室、办公室，甚至各种活动室和餐厅。丰富灵活的空间布局，不仅能够适应现代的个性化教学的需求，而且比起单调平直的长廊式的空间，更能引导学生的思维向多样化发展。

在日本和香港新建的高校教学建筑中，可以看到其室内、室外安排了许多适合停留、休憩、谈话的场所、空间和角落，并摆放有舒适的椅子或沙发；有条件的地方，还设置自动售货机、咖啡座、小卖部等等。屋顶平台的利用、走廊空间的丰富、门厅和中庭增加休息处等等都是创造良好交流空间的有效手段。

现代工业中心由于功能的高度复合性，面积达到了近8万m^2，如何在这样一组庞大的体

现代工业训练中心效果图

量中寻求空间的场所感是设计的重中之重，我们为了追求空间的群组感与聚合感，运用构成派的设计手法，空间中或是光洁的实墙、或是虚化玻璃体，或是片片构架，均成为空间中的重要构图元素，四个主要形体通过一个亦虚亦实的圆形，有机地形成了现代工业中心的核心交往空间，同时每个独立体量的交往空间体系有机的和核心体发生确定的或是不确定的关联，组合成了整个现代工业中心的生动有机的交往空间体系。

现代工业中心特别注重了交往空间的营造，同时需要明确的一点是，这里所谈的交往空间不仅仅是我们传统意义上的建筑空间，而更重要的是建筑空间所传达的精神空间——场所精神。在马斯洛的需求五层次论中，把人的需求分为生理、安全、人际交往、尊重、自我实现五个层次，这五种层次是递增的，满足了低层次，必然产生高层次的欲望。我们的社会，我们的建筑，我们的大学校园也是这样。在2001年之前，我们的校园建筑采用传统的现代主义的手法，大量建造了一批批量式、流水线产品后，已基本解决了使用、功能、安全的生理问题，但是站在今天来论述，我们提出对建筑空间的精神层面的更深入的追求——注重人与人交往的场所空间。

建筑格局的多元合成

科学技术的进步产生了整体化、集约化的趋势。在新的技术革命的冲击下，教学建筑中采用的各种技术设备需要管道、管井等，整体化、集约化的教学楼布局相对于分散式独栋教学楼布局在设备管道的利用率等方面均能较好地满足要求。速度和效率是信息时代的特征。除了完善的通讯网络提供工作的速度和效率外，建筑本身也要求能够联络方便，同时，建筑群也以成组成团的方式组合，尽量减少间距及交通路线。各个相对独立的区域之间，尽量打通分割的界限，加强联系。设计通过通道和连廊，使建筑群在整体上能够联络通畅，达到提高和保证交往、交流、传递和沟通的最佳效率。

如今校园与社会的联系比以往更加密切。许多高校都在城市内部或郊区，由于用地资源的不足，也希望学习空间的集中、便捷与灵活，这是教学建筑由松散的单体分散式的布局趋于向集约化的综合体格局发展的原因之一。此外，教学建筑布局的高度集中，在建筑用地指标不变的情况下，为扩大室外空间甚至高校整体校园外部环境提供了可能。

当前，随着经济、社会、文化的不断发展，为适应新形势下学习、生活方式的转变，反映独特的校园风貌特征。必须认识到一所学校的整体风貌和形象不是由一幢两幢建筑来决定；而是在协调好建筑物之间的相互关系，细腻而独特的外部环境及构成元素——建筑的外部形态的前提下，才能表达其独特的场所特色。现代工业中心的设计进行了详尽的调研工作，对各空间要素进行了理性分析，提出了最基本、最重要的依据，既创造清晰的虚实结构，完美的场所空间，又营造出独具特色的学校环境。

从上海工程技术大学的总图分析，入口至图书馆有500m，主干道两侧尺度较好，但缺乏收放有致的广场交流空间，我们在主轴的末端营造一个强有力的形态中心，但她必须内敛雅致，有向心性，这是现代工业中心圆形的设计由来。

这种围合的形态中心的原形可以追溯到希腊文明时代，正像神奇的雅典人聚会在一起畅谈他们的理想和对理性的敬意一样，现代工业中心这种非对称的圆形空间围合的形式，不仅提供了交流的场所，也传达着一种自由开放的信息，是一种不知不觉的共享，一份内心情态宣泄的舞台。圆形的形态中心处于现代工业中心的核心位置，成为日常学习和生活的延伸，社会功利的一面在此庄重的人性色彩中获得稀释，同时，尺度为直径60m的非对称圆以一种向四周通透开放的姿态，欢迎学生参与到这个环境中来。

抬高的圆形广场发挥着主导现代工业中心气氛的重要作用，向每个人提供教学楼的整体意象。在构筑中心边界的设计手法中，采用了建筑非对称限定，地面高差限定和交通方式限定等多种手法。设定这个形态中心时，环境因素越丰富，圆形的结构化倾向就会越明显，也就越有凝聚力和向心力。

现代工业中心的设计，把复合单元的边界围合而构筑出的非对称圆形的必然的空间形态，以旧典新唱的巧妙构思，把空间中或是光洁的实墙、或是虚化玻璃体，或是片片构架，把高雅的空间情节和潇洒、精致的现代手法融为一体。强调空间的语法、句法和词汇的永久生命力，揭示的正是这种作为有序的、易解的和共享空间建筑的最基本意义。

至此，也让我们可以深刻理解文艺复兴油画《泉》的作者——用生命的线条留下卓绝的千古素描——安格尔的名言“圆形是一切形中最至真至美的图形”的深刻寓意。

我们以上海工程技术大学松江新校区现代工业训练中心的设计为契机，对现代高校教学建筑的设计发展作了有益探索：现代学科发展的趋向，“交叉”、“联系”、“共用”等行为的加强，要求高校的教学建筑趋向集约化；尤其近当代科学所面临的各种课题的复杂性，要求现代和未来的科学工作者不能限于个人的研究活动，而应与他人全方位地交流和协作。集约化的高校教学建筑，更有利于不同系科师生之间方便、快捷的交流与沟通，从而开阔师生的思路，启迪边缘学科的开拓。因此可以预见，高度集约化的建筑布局和开放型的建筑环境逐渐成为新型教学建筑的主流走势。

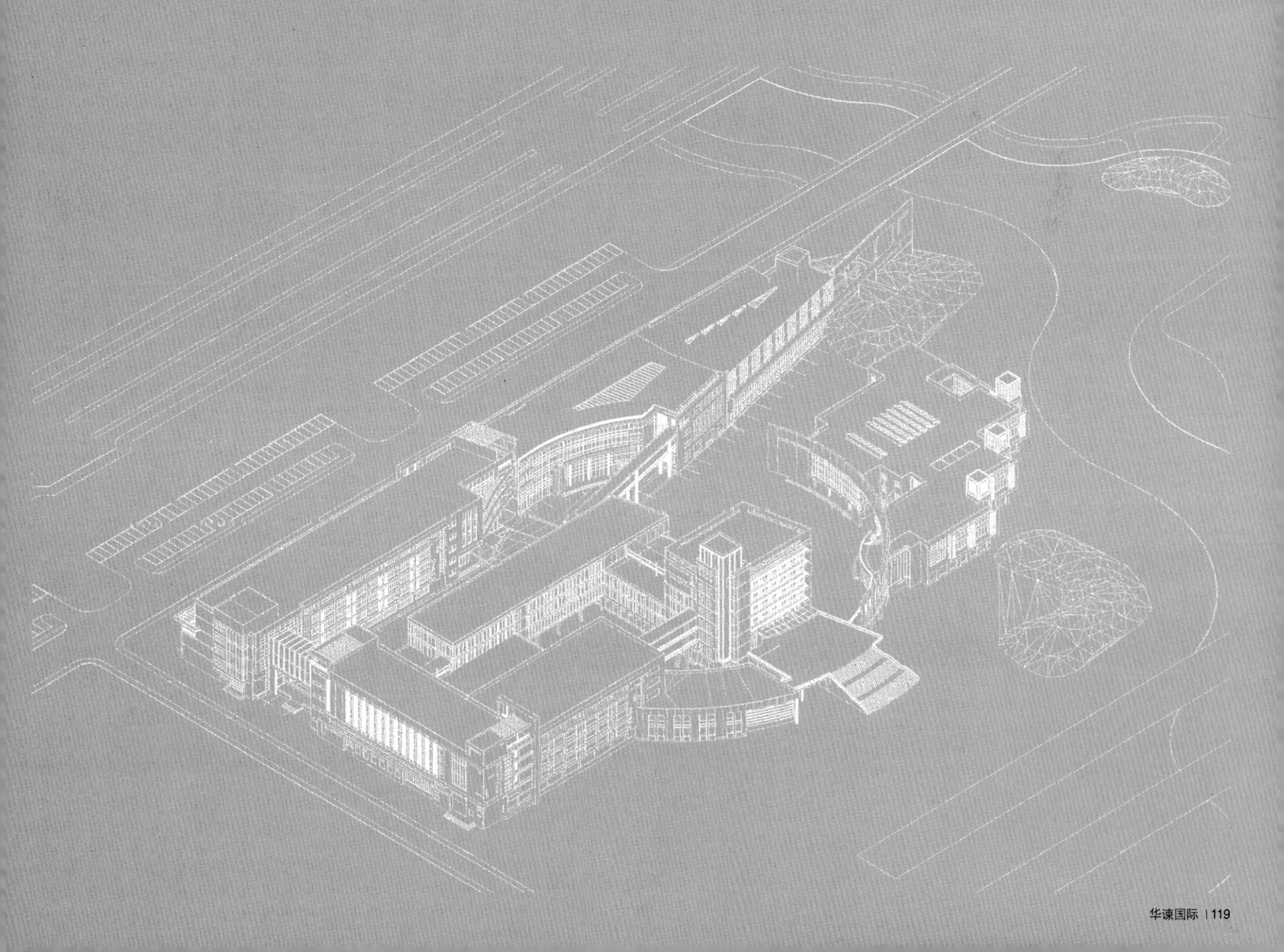

Landscape Design of Minhang Campus, East China Normal University

华东师范大学闵行校区总体景观设计

Shanghai, China, 2004.

华东师范大学闵行新校区2002年全面开始建设，校园建设量非常大，有众多设计单位参与其中。但除去总体规划对校园的整体控制，没有对新建设的校园进行更为深入的系统整合。在这样的大形势下，针对校区设计与建设的不和谐，学校决策层提出了新校区总体景观设计的要求。通过校园景观系统的设计与建设，校区入口等形象工程的精心设计与建造，增强校园风貌的整体性并完整的呈现给世人。

华东师范大学闵行校区景观设计是对校区规划设计的完善与发展，设计在深刻理解校区总体规划的基础上系统展开，综合运用轴线、曲线空间序列、折线空间序列等总体设计手法构筑起特色鲜明的校园景观框架。

整体景观设计注重对自然环境的合理利用，特别是江南水乡地貌特征在景观环境中得到很好的保留与升华，并将人工景观与自然环境充分融合。既有江南水乡的地域性格，又融入现代景观的各类造型手法，同时引入艺术设计的一些门类，充分满足了现代人的个性化、多样性和生活品位多元化等多方面的需求，共同构筑了一片师大人心灵归属的净土。

设计着重对校区的主入口空间，主校门、形象主轴线、核心景观区、滨水活动区等景观建筑区域进行系统设计，特别注重建筑、规划、景观三者之间的联系，将校园相对多元的建筑形态整合到一套完整的景观空间体系中，同时对几个入口的建筑形式语言进行了深入的探讨与研究。

01 由外而内层次分明的校园主轴线景观空间。
The main axis with different layers from outside to inside on campus.

The landscape design of ECNU Minhang Campus will develop and complement the construction plan of the new campus, which constructs unique campus scenery by using methods such as spatial series of axis, curve and fold lines, based on the understanding of the master plan of ECNU.

The landscape design pays much attention to making good use of natural environment, especially the character of southern China watertown being well preserved and improved, while natural environment being fully combined with man-made environment. The design contains character of local watertown, various methods of modern landscape design and several kinds of art design so as to meet people's needs of pursuing individuation, variety and multi-oriented life, all of which construct a land belongs to the spirit of faculties and students in ECNU. The design focuses on the area of landscaping buildings including main entrance space, main gate, main axis, main scenery and water platforms, especially the linkage between buildings, planning and landscape. The design tries to integrate a relative diversified architectural modality into a whole landscaping space system and to demonstrate the further study of the architectural language for several entrances.

02

02 校园的主体景观构架是一个以图文信息中心为支点的十字形，“十”字的各边功能各有区别又相互补充。
The landscape frame of campus is cross based on the information centre
Different functions on different sides if the complement each other.

03

04

03 04 校区主轴线景观设计不仅要强调空间序列上的层次变化，同时更需要强调不同时空场景的营造。

The landscape design of the main axis not only emphasizes the layer transformation in space but also creates comfortable surroundings in different time and place.

05 极具浪漫主义色彩的主入口夜景。

Romantic night view of the main entrance.

06 气势宏大的主入口广场及雕塑化的入口建筑。

The grand entrance square and its buildings with sculpture tendency.

基本资料

地理位置： 上海市闵行区

用地面积： 1,000,000m²

景观设计面积：699,420 m²

驳岸长度： 6200 m

容积率： 0.51

绿化率： 39.02%

设计时间： 2004年

BASIC INFORMATION

Location: Minghang District,Shanghai

Base area: 1,000,000m²

Building area: 699,420m²

Bank length: 6200m

Plot ratio: 0.51

Green ratio: 39.02%

Time: 2004

05

06

07 景观逐渐形成的华师大校园。

The scene on campus is going to be finished.

07

从前门到后门的故事

——华师大闵行校区景观设计

如果说建筑师的建筑设计追求形式、比例的完美，追求建筑语汇的准确表达，表现自我对于建筑风格的理解，表现对于美的领悟，是一种理性的书写；那么，当建筑师遇见景观设计，则给了他或她一个情感抒发的契机。

当设计的过程已成往事，留在心头的难道还只是某一条景观轴线设置？某一个轴线对景的安排？某一片硬地与草坪的互换？

也许是源于建筑师对于久违的大学生活的依恋，也许是源于建筑师对于华东师范大学某种特殊的情感，也许是源于建筑师对于自然与浪漫的向往，这一次，设计编织成了一个故事、一篇散文、一首诗。

这里是上海紫竹科学园区，八年后。

就像北大有个未名湖，剑桥有座难别的康桥，华师大中山北路校区有条深情的丽娃河一样，师大闵行校区有片传说中旖旎的十洲云水。这片水，源自图文中心，流淌过景观主轴，最后缠绵于整个校区，倒是应了朱熹的诗句，“问渠那得清如许，为有源头活水来”。整个校园由此分为东西两区，轴东主文，轴西主理，照现在的话说，魔鬼与恐龙分庭抗礼。浪漫多元的曲线景观轴与几何规则的折线景观轴，汇于景观主轴的同一点上，成就了一段天作之合的佳话。

一进东川路主校门，一条笔直的林荫大道直扑校北，中途在图文中心上跳跃升华了一下，顺势滑向水中的那颗小岛，淡入了林中。和所有的建筑思路一样，这是一条门面路，所以法桐高大，银杏绚烂，还有白玉兰的壮美，再加上叠落的浅水面与跳跃的喷泉，和着一道银光，落入水中，渲染出了十洲云水那片传说中氤氲的开始。或许凡尔赛的景致也不过如此吧！

大四那年交完论文稿出来，雨下得正急，撑了把伞心绪乱乱地四下走，走到了这条静默的路上。一抬头，迎面不知何时一个高高的男人，赤着肩膊，只穿了件白色汗背心，左臂上抱了一个白衬衣蒙着的娃娃，不紧不慢地走着。我一时并未看清孩子身上蒙的是男人的衣服，只奇怪这样气质的一个男人怎么会如此装束就出来行走。等到走近了看清了，心里莫名的一阵温暖。擦肩而过时，孩子的笑竟跳过雨丝咯咯而来，与男人挂在侧面的笑意一起，朦胧的主轴线广场清晰了，我的心绪平静了。

这不期而遇的温暖让我的怀念从这条内涵丰富的路开始延伸。

进了校门，就是国旗广场，左侧靠墙的草坪，据说不久要为新生建筑取代。国旗广场北头直接十洲云水的楔子。毕业时全班61名男生女生齐聚此地，听着浅水跃入河水中的铮铮誓言。记忆锁定为三个相拥而泣的女孩，红润的双唇在夜的背景与水面中幽蓝的灯光前，如盛开的花瓣，永不褪色地宣示着我们的青春。

向前，退离大气磅礴的浅水池两旁的，是两方巨大的草坡．这草坡是情人的席子，娇黄的蒲公英便是他们开开落落的誓言，秋日来时，皱缩的誓言便撑着小伞，悠悠地随风飘散了。夏日的草坪，每一片叶尖上都悬着一粒水珍珠，或许是浅水池里的喷泉开了小差．阳光将水池整齐的虹彩铺在草坪上，再带着露珠缓缓收回。中午无人，草坪是几棵香樟的，香樟的新叶是淡黄色的，半透明的，阳光一照，一树的明亮。那几棵樟树常常被男生用来作椅背，即兴的朗诵，或许会打动个漂亮的MM，故事便会转到夜幕下继续上演。

西校门不能不回忆：一道飞桥，跃过十洲云水；进了西校门，便是校区的景观次轴。西校门最美便是清晨，缓缓的云水推摇着惺忪睡眼的莲花，晨雾在这里萦绕，在这里留下个“清露晨流”的美名。次轴的对景依然是“源”般的图文中心，交叠起落的坡地与小径，倒真是镌刻着的流水，坡地草色的葱郁，不由让人再次咀嚼着朱老夫子那段“清”与“源”的哲学思辨了。最爱一旁的木栈道，一折又回到了主轴，由此引出的是师大最动人的银杏林。全国的大学生们都知道这么一句话："吃在……，住在……，玩在……，爱在华师大。"前面的都忘了，最后一句刻骨铭心。也许别处还有不少应景的盗版说法，但只有华师大，每一片草叶都是为了河与水的爱情而生。假如我是一只荆棘鸟，我就会选择栖在师大的荆棘上，把脚下素洁的白玉兰，唱成王子手中深情的红玫瑰。银杏的清香里，我眺望远处的云水最开阔的一段，还有那颗有着好听名字的小岛。人们都说银杏树又叫公孙树，树龄悠长，不知它们会把我的故事记录在哪一圈年轮里。

出了银杏林，眼前为之一阔，一道白练铺了开来，模糊了我的眼睛，其中那颗有着好听名字的小岛，清晰了起来："月波凝滴"。

樱花该开了，三月正是看花的季节。我一直也不明白，樱花的绽放为何那般热烈，那般繁华，那般雍容，那般凄美。也许因为有了极致的美丽，凋谢时才特别地令人伤感，令人手足无措。樱花的凋谢如雨，不是一片一片的飘落，风一过，纷纷而坠，铺起一地的眼泪。我曾接了一裙兜的花瓣，幻想着回去装一个软软的枕头，梦也变得香软。可是我忘了，樱花会让人深深伤感，一沾上，就是一生的挂念。

还有绣球，绿叶如玉，细碎的花白得极为贞节，花间梁祝翻飞。

海棠呢？朱红如血的铁梗海棠，粉嫩娇羞的垂丝海棠，枝干遒壮的木瓜海棠，还有未曾谋面的西府海棠。夹竹桃也是季节了吧，临河自顾，一树一树的雪白粉红，佳人临镜般的顾盼传情。江南的雨季，该绽放着怎样欲滴的娇艳。

梦回无数。

脚下站着的，层层淡入水中阶梯状的花簇，是叫作"秋色涟波"。还记得曾有男生指着图文中心酒后狂呼："我要以诗为梯，爬上你顶楼的窗台"

有趣的是，云水的对岸，张拉膜结构下的水中舞台，向北舒展出阶梯状广场，如同山水画中那晕开了的墨迹，晕出了个圆满的名字："烟波满月"。

也是圣诞节，星星一样的路灯把整个校园点缀得像童话中的宫殿一般，轻柔的乐曲漫空飘扬，十洲云水在沉沉的夜色中泛着幽幽的光，三十多堆篝火，映红了每一寸记忆。那些经冬不凋的冬青，那些蒙霜不枯的小草，那些闪着瑰丽梦幻的星星，那些铮铮的誓言，那些饱满的信心……忘忧的年龄啊！从南到北，云水的宽阔处，书写着"秋色涟波"、"月波凝滴"、"烟波满月"的主轴三景，总是让人不禁想起"长江后浪推前浪"的时代动力。或许，在主轴延伸的末端，站在"烟柳长亭"之上，三遇十洲云水之时，看着那如曼如歌的烟柳舞动，才会明白豪壮与浪漫息息相关的脉动.

再向北，就不能不走到宿舍区，那临水三叠的木栈道前，云水中晃动着"碎影斜阳"，还有怀念的思绪，漫无边际了。你也像我一样怀念过吗？心碎的感觉。再向北，就出了校门。

回头吧。顺着云水，向东南，关于食堂的故事，就不尽其数了。还记得食堂前的那个水坝吗？贴水的木浮筑外，瀑布如虹，藏匿于其中，倒有了几分世外桃源的幻象。隔着水幕看师大，难怪有着"画屏天畔"的美名！

走得再远，还是要回头的，再走一遍。站在"烟波满月"处，向南回眸，咀嚼着"满"字的韵味，看着东侧主文而去的浪漫多元的曲线景观轴，心绪也如草书般遒然苍劲飞白在校园的东侧。步移景异，移步换景的园林感受，难道是到了梦中的江南园林？醒来处，却在文学院的内庭。桂花的清香、红枫的艳丽、紫竹的清翠间散散点缀着几处小石凳，悬腕写出的曲线留下一笔白砂石，正是流水的影子，却有谈天论地的豪迈，"曲水流觞"，唱着千年的文化。

顺着主理而去的几何规则的折线景观轴，抽象与隐喻，穿梭跳跃在忽放忽收的理性思维中，严谨的西方美学观，将排列与秩序幻化作另一种浪漫，书本般折起的草甸下，总能寻着数理逻辑的影子，或许理学院教学楼间那片"书田阡陌"，可能教会你一生的道理。

从前门到后门，像一段按了循环播放的录像，在我的生命里演播不息。前门到后门，一步又一步的丈量，需要走一生的距离。

景观设计就是一个故事、一篇散文、一首诗。它源于生活的体会，它源于情感的沉淀，它源于对细节的感知，它源于一个未曾讲完的故事。

SIIC Chongming Dongtan Training Centre

上海实业集团崇明东滩培训基地

Shanghai, China, 2003.

上海实业集团崇明东滩培训基地的改扩建工程是上海实业集团崇明岛开发项目的一部分，包括培训基地办公楼和休闲住宿中心两部分。项目地处上海市崇明区东滩开发区内。

崇明岛是我国第三大岛，是世界上最大的河口冲积岛屿，地势平坦，每年还有数以万计的候鸟将这里作为迁徙的必经之地。项目所在区位——东滩，是世界级湿地自然保护区。由于当地脱离大陆，自然环境基本未被干扰和破坏。

培训基地采用典型的建筑表皮处理手法，将基地内的新建建筑与原有建筑有机地交合在同一个“建筑盒子”中，新建部分着重强调了对建筑外部的形式语言和建筑内部的空间语汇的整合与变奏，通过体与面的变化来表达新建建筑的时代特征；原有建筑在保留其特有的结构体系之外，在表皮上借用新设计的建筑的空间界面，联系中强调其平实的个性特征。

设计的过程特别像是在做一个“形象包装”——新与旧的问题、变与不变的问题、突破与延续的问题——集中体现在这样一个独特的个体当中。

01 上海实业集团崇明东滩培训基地是上海实业集团开发崇明岛的原点，设计之初就确定了最为生态和节能的开发方式，该图反映的建筑形象是对原老建筑改造后的成果，运用钢构与木构组合的复合立面正是开发策略最好的体现。

Dongtan Training Centre is the first project of SIIC to develop Chongming Island. Ecology and energy are the most important two elements concerned in design. The building in this picture is renewed from the old one. The mixture of wood and steel in facade is the best way to follow the developing directions.

02 基地改造前后组照。

Buildings before and after reconstruction.

基本资料

地理位置：　上海市崇明区崇明岛
用地面积：　90,000 m²
容积率：　0.10
绿化率：　75%
新建办公楼：
建筑面积：　3,136 m²
改建办公楼：
建筑面积：　2,166m²
改建食堂：
建筑面积：　431 m²
新建住宿接待中心：
建筑面积：　3,101 m²
设计时间：　2003年

BASIC INFORMATION

Location: ChongMing Island, ShangHai
Base area: 90,000 m²
Plot ratio: 0.10
Green ratio: 75%
New office building
Building area: 3,136 m²
Renewed office building
Building area: 2,166 m²
Renewed canteen
Building area: 431 m²
New hostel
Building area: 3,101 m²
Time: 2003

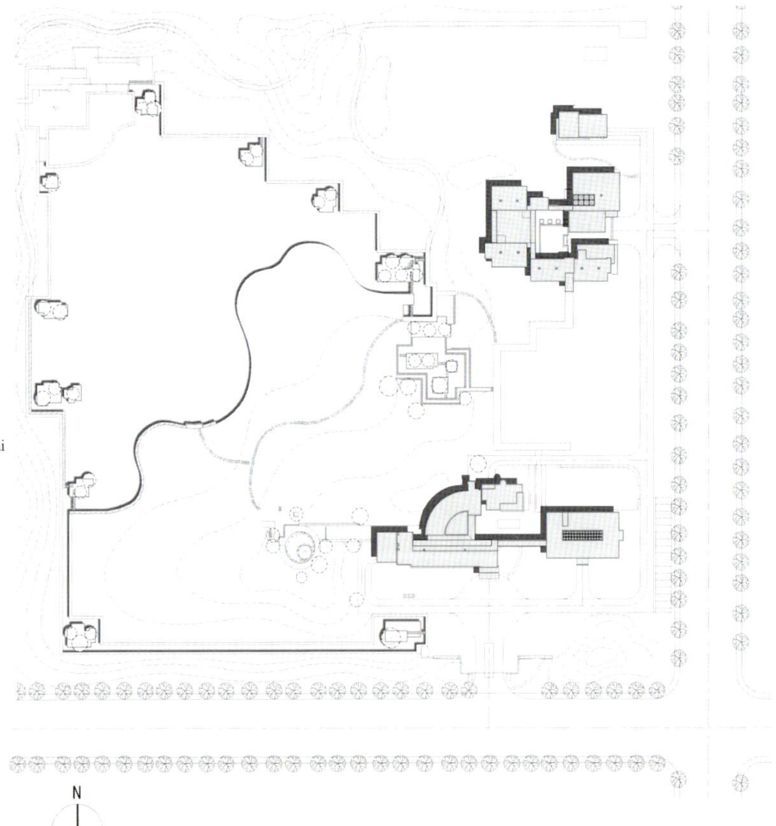

Site plan / 总平面

03 入口右侧的实体建筑为原三层高的老建筑，设计依据特色的改造策略将功能性要求更高的办公楼主入口设置在二栋楼的中间，运用虚实对比的手法将入口与两侧建筑融合为一个相互协调的整体，立面层次分明，节奏轻快。
The buildings on right side of the entrance are the original three-story office buildings. According to the plan, We put the entrance hall that needs a higher demanding in the middle of them. The three buildings are well integrated by the contrast between void and solidity Thus, the facade looks clear and light.

SIIC Chongming Dongtan Training Center is part of the developing programs of Chongming Island facilitated by Shanghai Industrial Investment (Holding) Co. Ltd (SIIC).

Chongming Island is the third largest island in China and the biggest estuary sedimentary island in the world. As a migratory base, Thousands upon thousands migratory birds go across here every year. Dongtan, where the training center stands, is a world-class everglade Natural Reservation with beautiful natural surroundings. In the training center, the original office building and canteen will be maintained and enlarged, while a brand new office building and a hostel will be set up.

In the design of original office building and refectory, a traditional method of architectural skin transformation is adopted to integrate the new and the original into a 'box', when the designer tries to regenerate the functions of old buildings. Enlightened by German Pavilion of Barcelona International Fair, the design of the new hostel emphasizes harmony and unification of inner and outer space on the basis of respecting and making use of the environment. Thus, Mies's flowing space reappear.

The process of design resembles an image decoration, which contains many conflicts including the old or the new, change or not change, variance or inheritance.

CHONGMINGDONGTANSHANGWUBANGONGJIDI2003_6
04

05

06

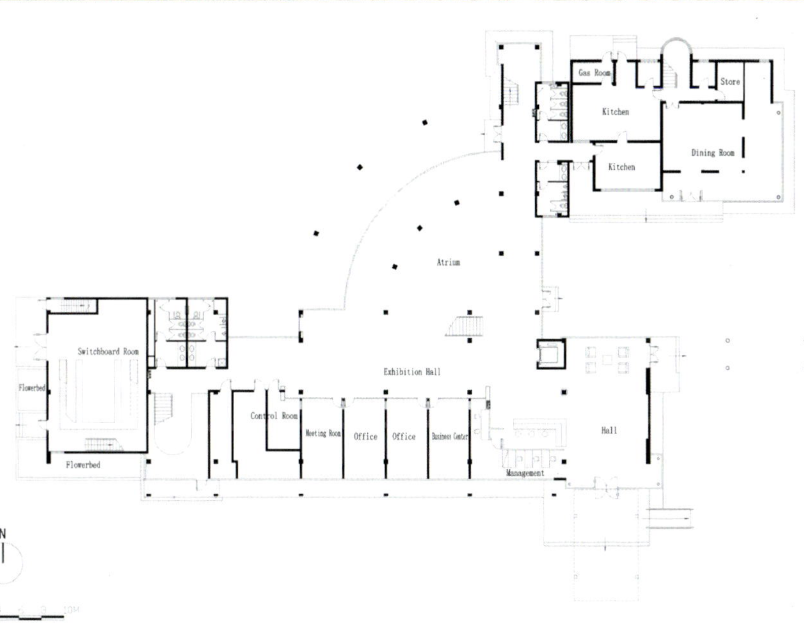

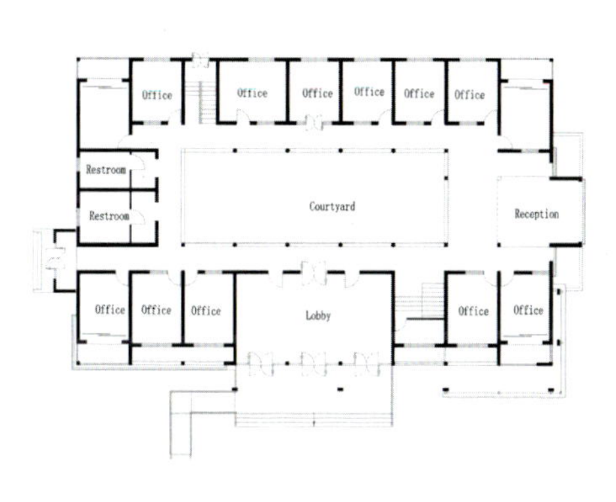

Fisrt floor / 办公楼一层平面

04 05 06 设计过程中的一组三维虚拟模型。
3D models in design.

07 设计手法更为多元的办公楼效果图。
The final design, using various techniques.

08 建筑东南侧实景照片。
View from southeast.

09 2层高的活动中心是住宿接待中心整体建筑的外延。
The two-story activity centre is a newly-built part affixed to the old one of the hostel.

10 设计手法简洁明快的住宿接待中心后勤服务楼。
The logistic house of the hostel, designed by simple technique.

11 玻璃体楼梯间，钢结构连廊，实墙面显现出三种建筑表情，时而轻灵，时而活泼，时而严肃。
The glass made stair house, steel-made corridor and brick made wall display three expressions of the building. Sometimes flexible, Sometimes active, Sometimes Serious.

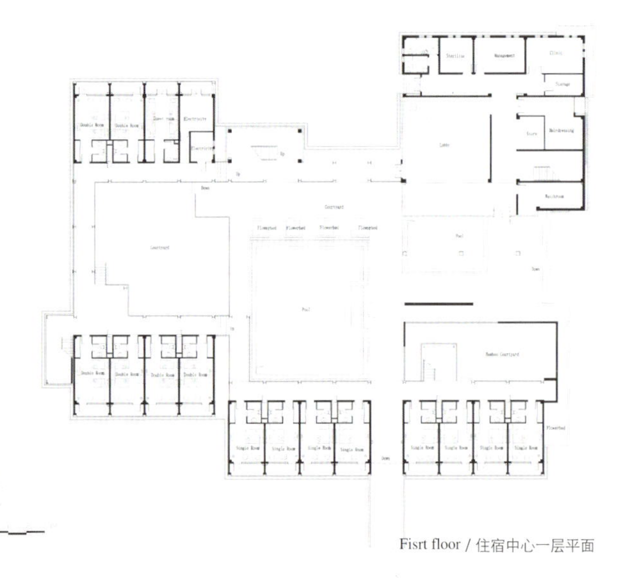

Fisrt floor / 住宿中心一层平面

⑫

⑬

⑭

15

设计之舞

上海实业集团崇明东滩培训基的设计思考

设计永远被林林总总自然的、人为的因素影响着，许多建筑师都把建筑设计比作戴着镣铐跳舞，这些限定设计思路的因素同时也就成为了我们设计灵感的来源。因此，设计的过程就成了一个反复定义问题和解决问题的过程。

定义问题

上海实业集团崇明东滩培训基地的改扩建一期工程是上海实业集团崇明岛开发项目的一部分，包括培训基地办公楼和休闲住宿中心两部分。项目地处上海市崇明县东滩开发区内。崇明岛是我国第三大岛，是世界上最大的河口冲积岛屿，地势平坦，每年还有数以万计的候鸟将这里作为迁徙的必经之地。项目所在区位棗东滩，是世界级湿地自然保护区。由于当地脱离大陆，自然环境基本未被干扰和破坏。

通过对现场的考察，虽然行色匆匆，但我们发现崇明当地建筑没有自身的特色，这是一个自然环境资源丰富而人文资源相对匮乏的地区。崇明地势平坦，临近海边，风很大，当地居民有在屋后种竹的传统。所以，要反映当地风土民情，我们选择气候和自然因素作为切入点。而环境直接、有力地影响着设计，不管是原有自然环境还是人为建设的环境，我们的目标是使这些因素在建设后重新达到平衡的状态。我秋季再到现场时更加坚定了这种想法。基地对面农田里一群白色的大鸟跟着犁田的翻土机吃虫，又大又笨的铁家伙和美丽的鸟儿亦步亦趋，相伴而行，使我确信人和自然是可以和谐相处的。

从项目开发者的角度来说，他们对开发定位已经有较清楚的认识了。业主给建筑的定位是"壮美的大花园"，"建筑要成为大地的艺术，旷野中的线条"，建筑要谦和，空间要丰富，"远看不起眼，近看有品位"，要做符合当地风土的现代建筑，并且形成崇明东滩今后开发项目的材料公约，在高度、体量、材料上形成建筑典范。但作为非专业人士，他们对如何达到这一目的，如何体现他们的想法，却没有完整清晰的认识，甚至会对一些设计手法带有偏见，而最初提供的设计要求也很难准确地反映他们的真实想法。这时"建筑师如果只是按照任务书的字面含义进行设计，往往很难真正地满足业主的要求。"

我们的首要工作就是定义这个设计问题。在通过几次沟通后我们基本明确了业主的想法，这样就基本产生了解决问题的办法。荷兰建筑师亨克·杜尔说："一个善于反省的建筑师应该善于倾听业主的声音，以批判的态度对待业主的观点，将那些观点翻译成建筑造型，并对整个社会持有批判的观点。"[2] 然而，建筑师和业主的沟通应该是无止境的，就像设计的完善可以无止境进行下去一样。在这个沟通的过程中，设计并不是沟通的唯一内容，积极了解业主的经历背景、知识结构、审美取向和价值观都是非常重要的。同时，在沟通中需要让业主对我们的专业产生理解，对我们形成信任感，这无疑会给后面的设计带来好的影响。

该项目设计建造的时间紧迫，业主要求整个项目从方案到建成在五个月内完成。所以，成熟的技术和构造方式成为首选，当地的地理地质因素在建筑的技术层面上起了重要作用。

解决问题

培训基地办公楼——新旧建筑的平衡

培训基地办公楼的用地非常宽松，西北面为大片人工水景和绿化，北面是休闲住宿中心的用地，而南面正对的是一望无际的田野。基地上已有两栋建筑，是当地团结沙农场20世纪90年代末建的办公楼和食堂，极具崇明岛"风格"，瓷砖贴面，女儿墙带着小翻檐。据筑曾是当地老百姓和农场领导的骄傲，建成后一直没有投入使用。业主出于经济因素和对当地领导和群众的尊重，要求保留原有建筑，只对立面进行改造，这既给设计带来了困难也同时带来了挑战。

功能组织上，我们以培训基地办公楼新建入口为中心，一边依次布置展厅、机房和变配电站，一边以庭院连廊和原有两栋建筑相连。建筑通过内外三个不同大小，相互渗透的空间组织起来。

设计过程中，曾在新楼中间形成一个扇形的生态庭院空间。庭院内种植乔灌木，布置水景，通过底层架空减少建筑进深加强自然通风，形成良好的小气候。但这一极富绿色意义的空间在施工过程中被业主否决，改建成了带采光顶的封闭展厅，实为建筑师最大的遗憾。

建筑南立面以透明的玻璃体，略突出入口，其余的体量大小基本相似，直线排开形成完整的一体。北向办公空间不同于正立面较严谨的直线关系，采用弧形体量立于开阔的草地上。既产生了多样的外部空间，丰富了造型，又凸现出建筑空间的张力，呼应了西北向的景观。弧形办公室每间角度都不相同，具有独一的景观，成为建筑内最富诗意的空间之一。

在立面上我们以水平线条为主，表现建筑依附于大地的态势。建筑用材主要是木材、钢、玻璃和部分石材。南立面的木构遮阳板既有实用意义又有很好的装饰效果。木构遮阳板的构造、颜色、间距在建筑师和施工单位之间反复讨论推敲，使得建筑的这一层表皮达到了审美、经济和技术的平衡统一。

面对原有建筑这一因素，我们立足于保持新建部分的主体地位，为不突显两者的差别，旧建筑立面改造采用新建部分的基本元素，在原立面外也局部安装水平遮阳板，加装了这一层表皮，并且玻璃、石材、钢构都保持与新建部分一致，最终，两者达到了统一平衡的效果。旧建筑面貌也焕然一新。

休闲住宿中心——经典空间的重构

设计构思的方法有很多，灵感的来源除了项目自身的各种因素外还有建筑师本身的修养。牛顿说，我之所以看得远，是因为我站在巨人的肩上。我们也常常从世界著名建筑大师的设计中看到前人著名的建筑元素。著名美国建筑师理查德·迈耶的设计手法成熟老到，作品堪称经典，而他的手法很多源自于早期大师勒·柯布西耶的创新。著名的施罗德住宅则是蒙特利安画作的立体表现。就连解构主义大师弗兰克·盖里离经叛道的作品中也有构成主义的影子。而他们之所以成为大师就在于他们善于继承经典，更在于他们对前人经典的进一步推敲和提炼并使之成为自己的经典。所以，经典建筑语汇的重构和重组是一种很好的设计手法。

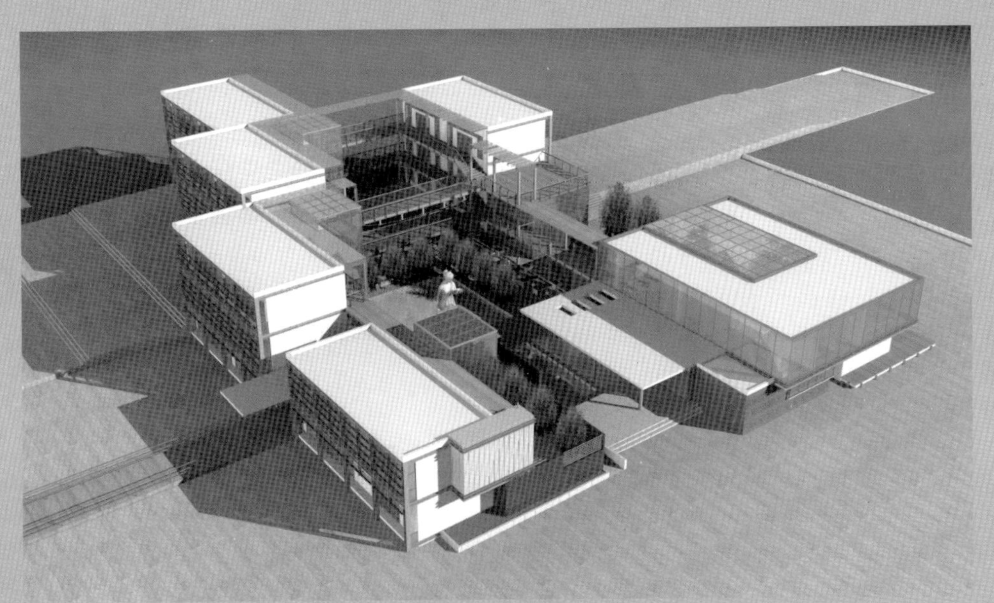

住宿接待中心三维模型

说，这两栋建在对休闲住宿中心设计要求的具体分析后，我们准备重构密斯·凡·德·罗的德国馆。该作品见诸于书刊杂志的常常是同样几个经典的空间，经过分析，我们把它归纳为渗透、流动、半围合、对景等几种空间类型，并在我们的设计中进行了重构。在建筑建成后我们希望能产生雅俗共赏的效果：从具有同样知识背景的建筑师眼里看到的是密斯馆的影子，而普通使用者感受到的是动人的流动渗透的空间。

我们再次遭遇了“传统与现代，国际与地域”这个时代性的话题。业主要求建筑要做成本土的、现代的、体现休闲气质的园林风格，同时又要节约成本加快进度。在这里，我们坚持认为继承中国传统在于继承传统建筑的精神而不在于一定要使用传统建筑形象的元素，而要体现时代精神就更是如此。在中国传统园林中的对景、障景和借景等手法形成了无数迷人的空间，堪称中国建筑的精华。所以我们决定采用钢、玻璃、木材等现代的材料建构建筑实体，用渗透、流动、半围合、对景等手法形成丰富的空间，同时照顾当地的气候和地理条件。

休闲住宿中心包括公共娱乐和客房两部分。在设计中我们将建筑水平铺开，各部分间用连廊连接，围合成了丰富的院落空间，充分符合休闲建筑的园林风格。公共娱乐部分有娱乐和服务功能，包括大堂总台，这一部分是我们设计的重点。沿基地的东面道路进入休闲住宿中心首先看到的是大堂入口，其雨篷出挑深远，类似密斯馆伸展的屋面。大堂向外的两面选用的都是透明白玻璃，透过这些玻璃，可以看到的是从建筑后部庭院渗透进来的景观。继续前行正对入口的庭院有景墙的引导，对面设置对景雕像，这一似曾相识的景象正是密斯馆流动空间的重构。在进入建筑和庭院的过程中，景观不断变化着，对景时隐时现，正体现了中国园林空间的“步移景异”的精髓。标准的宾馆客房具有成熟而确定的模式，为了节约设计施工的时间，客房部分我们设计为四个相同的单元。高级客房部分北面有独立竹院，符合当地气候和风俗，院墙采用磨砂玻璃，竹影婆娑，透在院墙上，造就了诗一般的意境。其余客房相互错落围合，形成高低不同的两个内部庭院，建筑的内部围合空间和外部旷野形成了有收有放的环境空间。

休闲住宿中心两部分的关系犹如中国画绘画技法中的写意和工笔。建筑只在重点部位细致刻画，其他部分一笔带过；空间主次有别，张弛有致。休闲住宿中心各部分间的连廊采用钢构框架，面层则铺以杉木地板。建筑底部架空，这既是为满足技术需要又增添了休闲建筑的气质。

设计后记

该项目于2003年12月中旬完工，历经6个多月，期间五易其稿。总的体会就是，建筑师除了埋头设计以外还需要关注许多其他的协调工作，可以说“建筑师的工作就是要不断地协调各种利益关系，最终找到一个大家都满意的方案。”(3) 而建筑师要实现自己的利益也就是设计的理想，需要通过不断的协商，运用知识和技巧最大限度地平衡各方面的需求，最终才能实现自己的利益。

注释：

(1)(2)梅卡诺与荷兰的现代建筑．方楠．世界建筑．2001（5）总第131期 第18页

(3) 善于反省的建筑师．亨克·杜尔．世界建筑．2001（5）总第131期 第21页

住宿接待中心实景

Tinglin Primary School

亭林小学

Shanghai, China, 2004.

亭林古镇已有1300年历史，是上海著名古镇之一。亭林小学作为一所创办于1905年的百年老校，其新校舍的设计是华谏国际在历史主义创作道路上的又一次有益尝试。

亭林小学是《上海2004中小学校建设标准》实施以来的第一个项目，在上海中小学建设史上具有划时代的意义。学校为全日制31班小学。规划设计采用行列式的布局，将体育运动场地设置在基地西侧，以一条贯通南北的建筑轴线连接校园建筑的各功能区块，由北向南依次为教学楼群、图书办公楼、实验楼、食堂及体育馆。轴线中心弧形部分是整个建筑群的中心，设置了办公用房和图书阅览室。教学楼分为四块，采用对称形式，由连廊巧妙相连。教学楼群和图书办公楼围合成半封闭的庭院空间，增强了建筑的场所感，创造出舒适的教学环境。实验教室设置在用地的南面，形成相对独立的空间。西南为体育馆和食堂，通过连廊通往各个部分。整个建筑群依据功能需要，结合基地形状和朝向，以不同尺度的开敞及封闭庭院组合，在较好地解决功能分区和交通流线问题的同时，也创造出一种"庭院深深"的书院意境。

建筑造型根据上海地方特点采用古典风格，三段式构图和古典线脚、壁柱等构图元素的运用，使建筑丰富而精致，体现了学校建筑的人文特质和历史感。办公、公共教学部分是整个造型的关键部分，庄重的弧墙弧廊，标志性的四个高塔，突出了建筑的视觉效果，表达了校园文化朝气蓬勃、积极向上的精神。

01 富有节奏感的建筑界面，奏出了希望的韵律。
The rhythmic interface of the building plays a promising cadence.

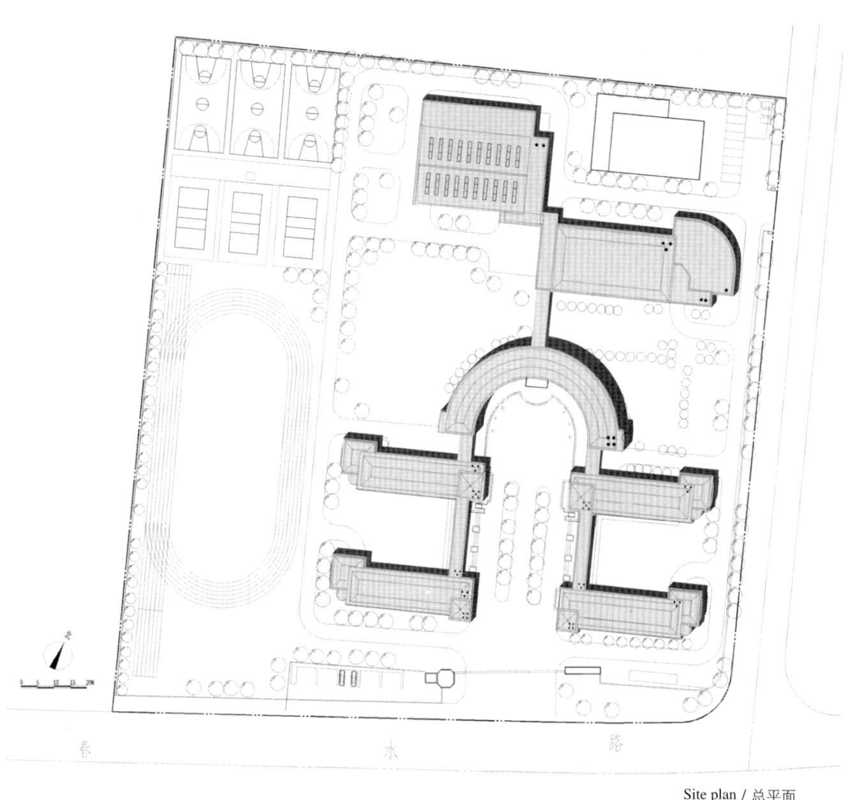

Site plan / 总平面

As one of the famous old town in Shanghai, Tinglin has a history of more than 1300 years. Tinglin Primary school was founded in1905 and its new campus design is a good attempt for DHGP .

The new campus of Tinglin Primary School is a milestone in school buildings in Shanghai, which was the first project when a local standard file of school construction was put into practice in 2004. The school is planned to build 31 classes. According to the plan, the designer is going to adopt the technique of the gradually progressed cortege parallelism. The playground is located in the west of the site. A south- north axis of buildings is planned to separate the school into several functional areas: the teaching building area, the library and office building area, the school refectory and the stadium. The arc in the middle part of this axis will become the center of this complex. The office buildings and public teaching buildings will be set up in there. All the classrooms will be partly, divided into four parts and ingeniously connected by corridors. The connecting corridors and the teaching buildings will form a half-open courtyard and a pleasant teaching environment. The laboratory is located in the southern part of the school, forming a relatively independent space. The gymnasium and the refectory are also in the southern part, all of which are connected by corridors. This plan meets the need of teaching and management, while making it pithy and sprightly taking the shape and orientation into consideration and solving the problems of the communication between all the functional sections.

The architectural style of the school is mixed with classical style according to the local characteristics of Shanghai. The decorousness, brightness, grandness and the accordance with the whole environment of the field represents the humanistic character of the school building. The plan is designed according to human nature. While different functional sections are connected with corridors. The office building area and public teaching area are the key part of the whole design. The solemn crooked walls and corridors, together with the four towers indicate the enthusiasm and persevering spirit of the school culture.

02 扇形的食堂与基地环境和谐相接。
The fan-shaped canteen connects harmoniously with the environment.

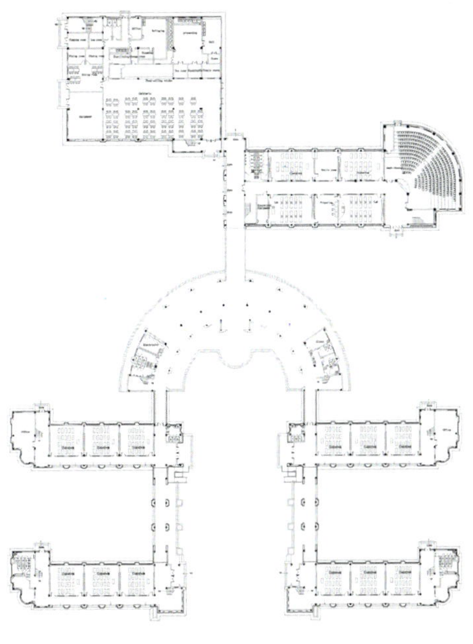

Fisrt floor / 一层平面

古镇·书院·新风

——亭林小学设计侧记

初闻亭林，就被这个颇有诗意的名字所吸引。然后就遐想这会是一个茂林修竹，枕石听风的悠然别处。

查阅书志，方知亭林乃江南名镇，境内文化胜迹颇多，明清启蒙思想家顾炎武（人称亭林先生）和亭林颇有渊源。而亭林小学，要追溯到1905年。这是光绪三十一年，风雨飘摇的晚清政府终于废除了延续千年的科举制度，全国各地新学堂纷纷设立，开启了中国现代教育的大门。同年，亭林小学的前身棗公立文义初等小学创立。算到今日，已逾百年。千年古镇，百年名校，历史文化底蕴如此深厚，新建筑将以怎样的语言展现其百年名校的身份并同周边的城市机理对话呢?

曾经的古镇，而今安在？所剩的也只是那诗意的名字和零落残破的遗迹，取而代之的已经是工业化时代的柯布式的新建筑了。历史的文脉已经被割断，这时如真出来一个青砖黛瓦、庭院深深的江南书院，反到像是豪华的赝品。

当真正开始上海金山区亭林小学建筑方案设计时，我的的确确采用了地道的新古典折衷主义手法：三段式的构图、长长的柱廊、高高的塔楼、券型的开窗、古典主义的线角壁柱……。待这个建筑群全部建成，我以嘉宾身份参加上海金山区亭林小学百年校庆的时候，这座设计精巧，造型优雅的校园放在亭林镇这个大环境之中竟是那样的合适，一点也不突兀，而建筑所散发出的含蓄内敛、蓬勃向上的气质显然使其已经成为整个街区当中标志性的景观，也成为亭林镇人民心中的圣地。我倒要仔细思考这所学校建筑风格设计的渊源了。

建筑总会反映出其所处时代的状况，从形式的选择到具体建筑问题的解决方法都可以看出社会经济和技术因素的影响。建筑还会反映出社会的文化取向以及自身文化的本性，因为建筑与文化的生命是紧密相连的整体。亭林小学选择新古典折衷主义的建筑风格也是和我国当代特殊的文化背景脱不开的。

当代的中国处在快速的现代化阶段，包括文化在内的各个方面全方位的在和国际接轨。现代文明是由以“两希”文明为基础的欧洲文化发展来的，而当这套现代文明推进到其它地区时，欧洲的文化就近乎是像标准一样强加到各色不同的文化上了，渗透到生活的各个方面，也包括建筑。

具体到本案，就不得不考虑到上海这一城市特殊性。在上海这个曾经的东方巴黎，作为中国最早接触西方文化的沿海城市之一，城市里有相当数量的新古典折衷主义的建筑，在文化上对于折衷主义风格具有较强的认同感。哲学家们对折衷美学思想作了概括，把种种经验描述归纳为一个基本概念，即强调建筑之所以成为审美对象，其根本的内在因素是建筑可以成为一种象征，而审美价值就在于恰当地、有效地象征出一种精神。而折衷主义的建筑风格在很多人的感情中恰是可以代表昔日辉煌的上海的。因此当新上海在重新回到世界舞台时，众多的新建筑纷纷采取折衷主义的语汇也就不足为奇，而且成为历史延续的必然 。

亭林镇作为上海郊区的一个现代化工业城镇，在城市化的进程中，选择折衷主义无疑可以更为快速的拉近与上海的距离，至少是在建筑形式上。古典折衷主义厚重的构图形式还可以告诉每一个参观者这是一所有着悠久历史的学校，并且还将在这片土地屹立更久。

就建筑领域而言，目前是中国现代建筑不断向前发展的时期，是一个探索时期。建筑风格上向多元化的方向探索，寻找一种能够表征自己文化的方式。纵观建筑的发展历史，探索时期经常是折衷主义美学思想最活跃的时期。我们的近邻日本和印度在形成具有自己文化特征的建筑风格之前都经历了这样一个折衷主义的探索阶段。折衷主义美学的价值又何在呢？“折衷主义美学是观念革新的产物，是对每一个阶段新时期建筑风格所不断探索的精神支柱。”建筑师通过对一个个实例更详细地分析研究，掌握了细部与建筑所处的总体尺度之间的关系，开窗形式对光影变化的影响，空间组织的序列变化，这都是古典建筑千年积淀的可怕魅力。

在建造折衷主义建筑的同时，文化的折衷像灵魂一样被输入其中。不论是选择何种地域何种时期的语汇进行折衷，本土文化是永远被包含其中的，在建筑的各个方面你依然可以发现其自身文化的影子，这就是建筑的地域性。全球化与地域性的问题是当今世界各国都在面对的问题。折衷主义下的全球化与地域性不再处于矛盾的对立面，而是将地域性变成了全球化的一个特征。

纵观中国文化史，每一次文明的融合之后都会带来新的文化高峰。诚如秦一统后的汉，唐开放后的宋。中华文化具有强大的包容性，总能以开放的精神汲取四方文化之精华，为自身文化注入新的血液，而繁衍至今。

就如同华谏国际（DHGP）设计的亭林小学，细味之下，即使是这样一个新古典折衷主义风格的建筑作品，也传递着越来越多的中国文化气息：一种由内而外的中国书院气质，一种来自建筑设计师所赋予她的性格特质。

含蓄内敛是中国文化的特征，层层的院落也可以看到中国建筑的影子，通透的连廊也似苏州园林那般步移景易。更为动人的是，从建筑的一窗一洞，一砖一瓦，我看到了华谏国际（DHGP）建筑师从一切优秀建筑形式中学习，对于中国现代建筑发展的研究，在现代建筑结构与古典建筑形式矛盾中寻找一种默契，使建筑带有一种怀旧的现代情节的创作思考。尽管这种革新尚不成熟，但表明我们已经开始了几千年来建筑史上的探索之路。

漫步校园，你是否看到了夏日墙上斑驳的树影？听到走廊里传来儿童奔跑的脚步回响？感觉到校园里那熟悉而亲切的空气的馨香？

鸟瞰效果图。

Shanghai BaoShan Weather Centre

上海市宝山气象中心

Shanghai, China, 2002.

上海市宝山气象中心新址位于上海市宝山区友谊西路南侧，江杨北路以东，由综合楼、氢气房、气象测量场等场地组成，是一处反映现代化气象服务和现代城市景观面貌的公共建筑。整体设计运用平面构图、空间变换和立面解构等手法来诠释科技化与人性化的气象文化建筑内涵，并将建筑的专业功能、科教功能、旅游功能与建筑的形式、空间、外部环境等和谐地统一起来。

主体建筑沿友谊路向后退让20m，形成气象中心办公楼门户空间，并与城市绿化带自然融合，结合主入口的功能特性，精心构筑了景观绿化广场，打破线性乔木植物空间模式，以草地为主，点缀小品式的盆栽树木。

气象中心综合楼是整个设计的主体，由科研、办公、会议以及后勤辅助四大功能块组成，平面形式别具匠心，采用巧妙的平面及空间组合手法处理四大功能块的组合，不同功能建筑空间的相互联结浑然天成。

立面材料选择了白色铝板与绿色镀膜玻璃，以纯净的材质凸显出建筑的科技感与文化感。空间与造型设计上强调建筑、环境与人三者间的对话，精致的挑板和轻盈的屋面交相辉映，圆弧与直线构成的趣味对比，塔楼与空廊在水平方向和垂直方向充满灵性的对话，入口灰空间与中庭空间的巧妙对话……整栋建筑犹如一个白色精灵，展示气象中心开放与包容的新姿态。

01 抽象的浑天仪，诠释了气象中心特殊的建筑性格。
The abstract armillary sphere indicates the special character of BaoShan Weather Center.

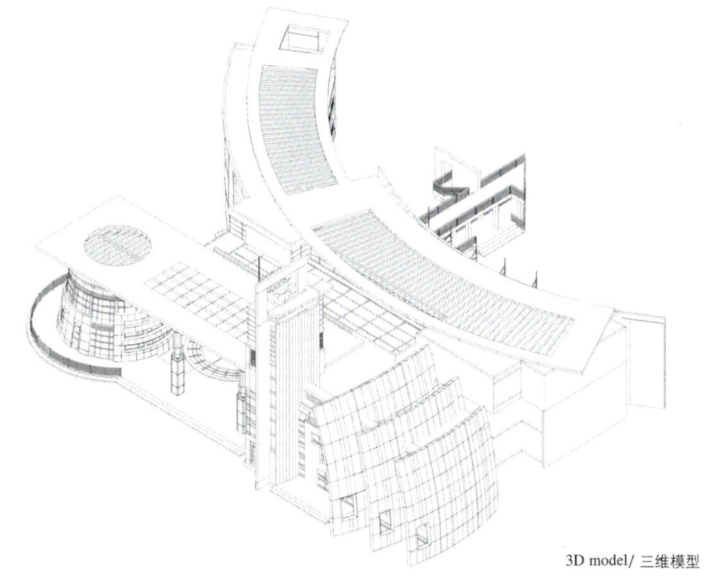

3D model/ 三维模型

02 建筑主展开面表达了气象中心的均衡美学，在水平与竖向的构图中插入三片弧墙，丰富了建筑空间层次戏剧性的变化。
The main facade of the building reflects balanced aesthetics of the weather centre. We inserted three arc walls in facade composition so as to add dramatic transformation in space.

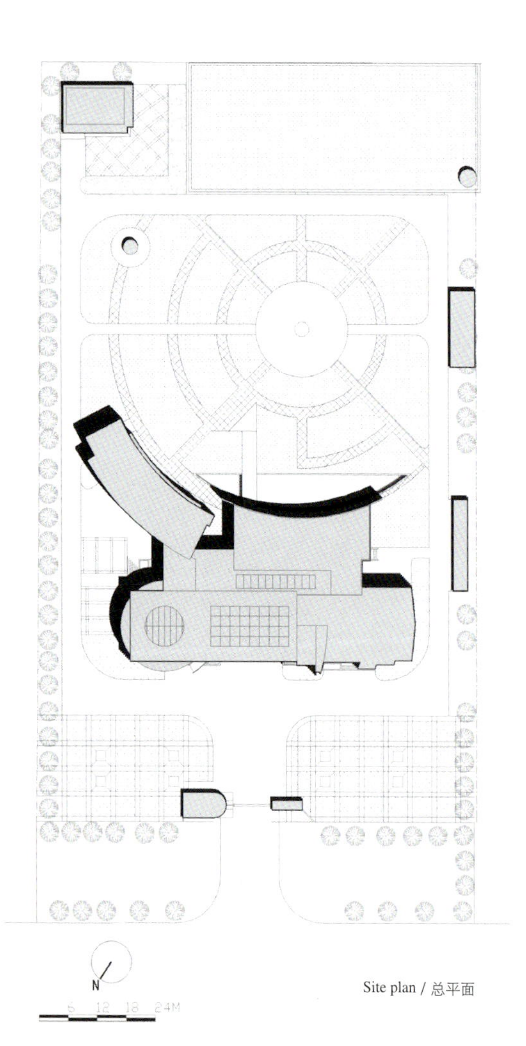

Site plan / 总平面

基本资料

地理位置：	上海市宝山区
用地面积：	13,546m²
建筑面积：	3,115m²
占地面积：	1,533m²
容积率：	0.23
绿化率：	61%
设计时间：	2002年

BASIC INFORMATION

Location:	Baoshan District, Shanghai
Base area:	13,546m²
Building area:	3,115m²
Site area:	1,533m²
Plot ratio:	0.23
Green ratio:	61%
Time:	2002

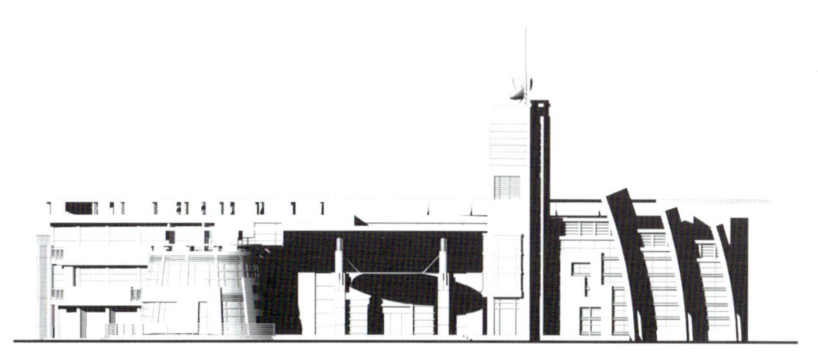

Shade / 阴影律动

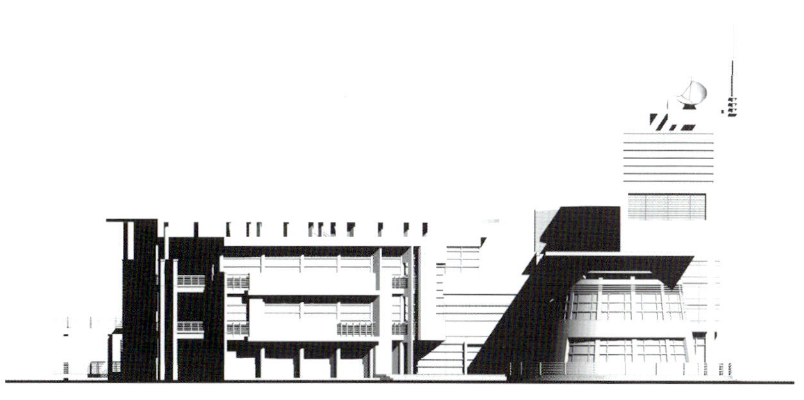

Shade / 阴影律动

The new Shanghai Baoshan Weather Center is located at the south side of West Youyi Road and east to North Jiangyang Road. The total site area is about 13,000 square meters. And its rectangular lot is a 150m x 80m piece of land.

The Baoshan Weather Center is a public building that reflects the modern weather service and our modern city view. To reflect those features, the designer tries to demonstrate high-tech as well as humanity via plan composition, space transformation, and facade de-construction. Thus the three primary functions (professional, educational and touristy) mixed harmoniously with the architectural form, space and environment.

The weather center falls back 20 meters from Youyi Road, forming the entrance square and green belt, which blends harmoniously with the city's green scenery. The green square is not a traditional one, where arbors are planted along the square edge. Instead, the designer decorates some potted plants on the grass.

Spaces for different purposes in the weather center are arranged perfectly. They exist well in the same plane and space but not disturb each other. People in different buildings may make the most use of the space and get the best view out of the window.

The design of the weather center is quite simple so as to get a modern look. The building indicates an atmosphere of culture and education. The surface and the roof combine perfectly. The arc and lines were cleverly decorated on the building. The weather center mixes well with the city and looks open and grand.

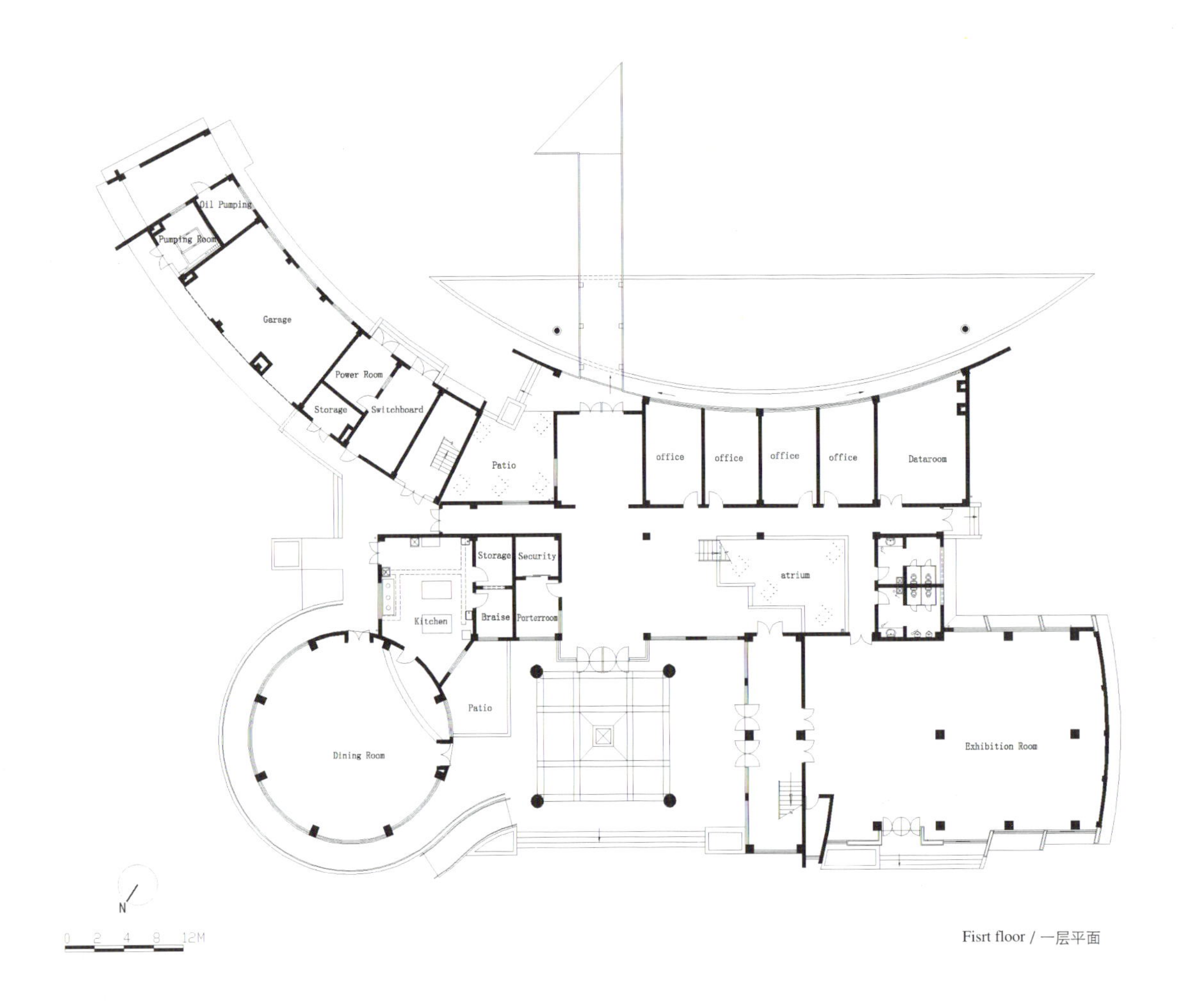

Fisrt floor / 一层平面

03 04 “浑天仪”不仅仅是建筑的一个特殊符号，更在不经意间使入口变得表情丰富，时而轻盈，时而稳定。

The armillary sphere is not only a special symbol of the building but also a changeful expression of the entrance. Sometimes it looks lightsome but sometimes looks stable.

05

06

在“浑天仪”上定格的镜头，表达了气象中心精神的、功能的、伦理的、文化的、历史的多重价值。

The close-up on the armillary sphere, which denotes multi-oriented values of the weather centre, including spirit, function, ethical, culture and history.

07

08

09

白色精灵

上海市宝山气象中心设计感悟

对话

人与人之间的对话，可以拉近两者心灵的距离；

建筑和建筑之间的对话，可以使得整个城市和谐统一、浑然一体；

建筑和基地之间的对话，可以展现基地隐含着的特性，最终建筑与场地相辅相成、融为一体；

光线和建筑之间的对话，可以赋予建筑灵魂；

而建筑和人之间的对话，能够使你的心灵与之共鸣

……

华谏国际（DHGP）在上海宝山气象中心设计中，秉承了追求最高境界的设计理念。垂直方向的塔和水平方向的廊之间的对话，让整个建筑不再是凝固的音乐，而成为跳动着的精灵；室内的人与室外景观的对话，使得工作者忘掉伏案的疲惫，换得身心轻松；入口的灰空间与建筑内共享空间的对话，使得从室外到室内的过渡显得如此自然，同时也体会到了中国古典园林中欲扬先抑的美妙之处。

看似只是细节上的对话，但对于华谏国际（DHGP）这样一个将“细节决定成败”作为座右铭的设计团体来说，这些对话是靠日积月累而形成的，并潜移默化地藏于设计师的笔锋。然而真正体现设计感悟的还是蕴含在其中的建筑文化之间的对话。保罗·里柯在《普世文明与民族文化》中曾说过这样一段话：“每个文化都不能抵御和吸收现代文明的冲击。这就是悖论所在：如何成为现代的而又回归源泉；如何复兴一个古老与昏睡的文明，而又参与普世的文明。”然而“这种相遇从未在真正对话的基础上进行过。这就是为何我们现在处于一种休止或中间休息的状态，在这种状态下我们无法实施单一真理的教条主义，而又无力战胜我们已陷入其中的怀疑论。我们处在一条隧道中，一头是教条主义的黄昏，另一头是真正对话的拂晓”。

“乱花渐欲迷人眼”，在这个各种文化，各种理念，各种思想并存的年代，我们一直为建筑的何去何从而苦恼；一直为是否要接受西方的先进建筑技术文化还是保留本土文化而犹豫徘徊，似乎如果选择了西方建筑文化那么就是忘本，就是崇洋媚外。有人不但无法接受自己的建筑师接受西方的先进文化，也不能容忍西方建筑师在“我们的地盘上”建造，甚至还有人大放厥词：与其让中国变成西方建筑师的试验场，还不如我们先下手为强。就像有人批评雷姆·库哈斯他不懂中国的历史却跑来建造劳民伤财的中央电视台。我们姑且不用操心他是否真的伤了中央电视台的财，就算是，那也是“周瑜打黄盖，一个愿打一个愿挨”。我们无法要求一个建筑要既经济又实用还要美观，或者更挑剔的还要求它在其中暗示某种哲学含义。就如同我们无法要求罗马的万神庙在满足了神的尺度之后还要适合人类居住，要求柯布那放之四海皆准的“机器住宅”要具有地域性；也如同我们无法要求赖特为富人设计的草原式住宅要为穷人服务。

对于华谏国际（DHGP）所构筑的宝山气象中心，这样一个地标性建筑，在满足了适用的前提下，最重要的便是反映时代精神。在这一点上，我作为设计师或许已经做到了将“借景”这种完全东方文化的设计理念与完全西方的“共享空间”巧妙结合进了一个设计中，不再为“向东走”还是“向西走”而犹豫，而是积极探索东西方文化的结合，以“取东西

的态度迎接“真正对话的拂晓”。

再谈理性与感性

从古至今，理性主义的魅力一直吸引着众多的建筑师在建筑创作中不断实践。对于古代理性主义的体现，如果追溯到文艺复兴时期，那么首当其冲的应该是建筑师帕拉第奥。帕拉第奥的理性体现在平面上“二与一”之间交错变化的韵律节奏。而现代建筑师中，能将理性体现得出神入化的当是柯布西耶，姑且不提他对古典理性主义的继承，就单是他自己独创的模数理论就足以证明他是理性主义的忠实信徒；即便如阿尔瓦·阿尔托这样以体现民族浪漫主义和有机主义建筑著称的建筑师，其建筑作品中也隐喻着斯堪的纳维亚的多立克理性。

而到了当代，理性主义概念涉及的面也越来越广泛。在我看来，理性主义还体现在建筑设计中分析问题，解决问题的过程。对于上海宝山区气象中心设计，面对的最大的问题应该是如何满足气象中心在气象观测方面的工艺要求。

建筑师有自己的解决方法：在总体规划上，将25.5mX60m的观测场布置于基地最南侧，标高较友谊路高40cm，结合绿化广场设计，广场上不种植高大乔木，以草坪为主，从而满足气象观测的要求。氢气房的位置满足了远离主楼50m的要求，放飞亭置于氢气房附近。三者皆处于办公楼值班室的视野之内，便于工作人员观察。在建筑单体设计上，主楼的一角设计了一座相当于5层楼高的雷达塔，塔的东南方向无建筑物遮挡，便于L-波段雷达的安装和使用，塔顶平台面积2mX3m。氢气房的层高6m，开高窗，通风良好，满足氢气房的防爆要求。氢气房开门3.5mX4m,朝内开启，汽车能直接进入，便于氢气瓶的运输及充气要求。在主体建筑上，气象专业的观测值班室和仪器室及酸雨室布置在一起，紧凑合理，满足气象工作人员的工作要求。

而理性是建立在感性的基础之上的。在建筑设计中，没有绝对的感性也就没有绝对的理性。假如说建筑设计有可能是绝对理性的，那么对于同一类型的建筑由同一个建筑师设计其结果即便不是完全相同，也不应该是大相径庭的。而柯布设计的朗香教堂与哈图雷特（La Tourette）修道院这两个同种类型的建筑，结果却是迥然不同的。所以，就连柯布这样一个对从“机器美学”理论，到模数理论、精确控制，对理性对数学如此执著的大师，也在日孚山区上创造了这样一个非古典力学、非欧几里德几何学的“朗香精灵”。而以经典历史主义设计风格见长的华谏国际（DHGP）也在友谊路上创造了轻盈舞动的“白色精灵”，我想这也应该是感性或者说是一个叫灵感的东西在作用吧。

从最初设计气象中心到最终建成，我脑海中总是浮现一个呼之欲出的形象：轻盈的舞者。外立面精致的挑板、轻盈的屋盖以及立面的白色材质，不正像一个身着白衣的精灵在轻盈地舞动吗？

景观设计之借景

古典园林中，有一种强化园林景深的方法，我们称之为“借景”。“借”就是把彼处的景物引入到此处来，这实质上无非是使人的视线能够越出有限的障碍，从这一空间直达另一空间或者更远的地方，从而丰富层次。借景分为远借邻借，旨在借园外景物弥补园中不足，这是十分讨巧的扩大空间与景域的做法，如：

水池不种荷花，留出水面反映白云、彩霞、明月，这是俯借；芭蕉、残荷听雨、“举杯邀明月”，则是应时而借；把远山、远塔引入视线，是为远借；做高视点俯瞰临园景色，是为临借。

按照以往的观点，建筑师大都在小区规划或者广场等设计中才会重视景观设计。实际上之精华弃其糟粕”对于办公建筑来说，景观设计也是非常重要的。面对周边的优美环境，设计师应该考虑如何能最大程度地将周边的优美环境引入到人的视线中来。所以在这个办公建筑中建筑师想挑战的可能不是如何将办公空间做得更具现代性，更富有机器美感，而是如何创造与环境结合的办公空间。当案头的文件堆成山，当被修改了无数次的计划仍返回来困扰你的时候，你的视线可以先从文案中抽取出来，移向窗外，这时大面积的玻璃窗和别致的观景阳台发挥了作用，窗外建筑师精心设计的景观被借进来。在这一刻，室外景观和室内空间交融了。

形体美学

一个时代有一个时代的审美意识。特定的审美意识又会诱导出符合及印证它的艺术形式。对老建筑来讲，形体美的法则是多样统一。而当代西方建筑艺术中正盛行着非和谐的、非完整的、非统一的形式。文丘里对于不和谐的建筑形象有自己详尽的主张：要“用不一般的方式和意外的观点看一般的东西”，“允许在设计上和形式上的不完善”。具体手法包括“不分主次的二元并列”，使“不同比例和尺度的东西”、“对立的和不相容的建筑元件”堆砌重叠。而形体美的法则在屈米那里则变成了解构、破碎、爆炸。他对这种法则的实现有拉维莱特公园为例。

这种审美意识从一个极端到另外一个极端的演变，可能是人们对长期无改变的一种美产生了“视觉疲劳”所导致的。然而屈米的解构美学并未坚持很久，在最近的雅典历史博物馆的设计竞赛中，他的获奖方案就算没有遵守古典美学的形体美原则，却也看不到丝毫先前解构的痕迹了。或许这又印证了一句非哲人所说的有哲理的话：历史是一个圆形的跑道，我们从起点到终点不过是转了一个圈。

华谏国际（DHGP）在宝山气象中心设计中运用的形体美原则最明显的便是体量对比手法。首先是水平方向的廊与垂直方向上的塔的方向性的对比，这是一种利用交替穿插改变各个体量的方向以求得良好效果的手法。而这种良好的效果便是得到了丰富的外部轮廓线，打破了方盒子轮廓线的单调感。再次是弧墙与直墙之间的曲直对比，直线的特点是明确、肯定，并给人以刚劲挺拔的感觉；曲线的特点是柔软、活泼而富有运动感，所以这种巧妙地运用直线与曲线的对比可以丰富建筑体形的变化。第三种便是入口灰空间与墙面之间的虚实对比。宝山气象中心的设计是将虚实与凹凸等双重关系结合在一起考虑，并能巧妙地交织成图案，不仅可以借虚实的对比而获得效果，而且还可以借凹凸的对比来丰富建筑体形的变化，从而增强了建筑的体积感。此外，凡是向外凸起或向内凹入的部分，在阳光的照射下，产生了光和影的变化，并且由于处理得当，产生了美妙的图案。

墙面与窗组织的形体美法则。在墙面的处理中，最简单但也是最单调的方法便是不顾内部空间，均匀的排列窗洞。在这个设计中，由于有科普展览室及值班室两个特殊功能而导致的一个竖向大空间和一个横向大空间的存在使得立面上采用大小窗结合的方式，这不仅反映了内部空间结构，还具有优美的韵律感。

说起形体美，我认为一种是形式上的，即看得见摸得着的；还有一种则是“有意味的形式”。亨利.摩尔有一段名言：“表现形式的美，和表现形式的感染力，两者的功能各异。前者取悦于感觉官能，但后者则带有一种精神力量，每能触动我们的心灵，实在更为动人。”恰是在此层面，宝山气象中心通过整体与细部的结合唤起了我们的审美情感。

建筑雕塑

雕塑　如果建筑也算是一个雕塑的话，那么在这个建筑中就有两个雕塑存在。一个是建筑本身，另一个雕塑是入口灰空间悬挂着的雕塑，我将之命名为“新版浑天仪”。我个人认为这两个雕塑在某种意义上说来并不能称之为雕塑。因为一般来讲，雕塑是静态的，而在这里，这两个雕塑是有灵魂的，是动态的。先说建筑这个大雕塑，它如同一座白衣的精灵，白色的遮阳板、轻盈的屋盖以及立面上三个突出的翼使得这个建筑如同在阳光下跳动的音符。而入口悬挂着的雕塑从形式上看，它是静止的，但是精心设计的天窗使得阳光透洒在雕塑“新版浑天仪”上，雕塑的影子随着一天中阳光的变化而幻化不止，此时此刻，此情此景，雕塑被阳光赋予了生命。此雕塑的存在还使得建筑性质得以准确诠释，标明了其气象类建筑的特征。

生态理念　在这个建筑中，我设计了一个边庭空间。冬天，这个空间是一个大暖房，是内部小空间的热缓冲层。在过渡季节，它是一个开敞空间，室内室外能保持良好的空气流通，有效改善室内的小气候。而到了夏季，由于有效的遮蔽直射阳光，边庭又是一个巨大的凉棚。在这个设计中，边庭空间的作用还不仅仅在于生态性的，由于它的存在使得大小空间交融，也强调了共享空间的感染力。另外，绿色植物引入室内，保持室内空气新鲜，改善室内湿度条件，调节室内温度。

空间感受　一个真正有灵魂的建筑，无论是光影变化还是虚实对比或是空间变幻都应该是无法单单用照片表达出来的，照片表达出来的只能是某一特定时间某一特定角度的没有生命力的形式。

参观者在这个建筑中移动，从入口灰空间到共享空间再到各个小空间，这种差异性的对比使得观者产生了情绪上的突变和快感。他在建筑中行走，就如同是在观赏一幅巨画，在聆听一曲长长的交响乐，视觉、听觉、触觉都被调动起来。最后，感官将建筑片断整合起来，建筑的整体形象在观者脑海中形成。而这些都是赋予建筑灵魂的必要条件，这种感觉可以套用柯布说过的一句话：“这是如此真实：建筑将被判是死还是生，取决于运动这一准则在多大程度上被漠视还是被非凡地开发利用。”

结语

如果让我以一句话来概括这个建筑的话，那么我愿意以“舞动的精灵”来形容它的本色。光线赋予建筑以灵魂，材质赋予灵魂以肉体，空间使观者与建筑产生心灵的共鸣，而三者结合的结果便是上海宝山区气象中心。

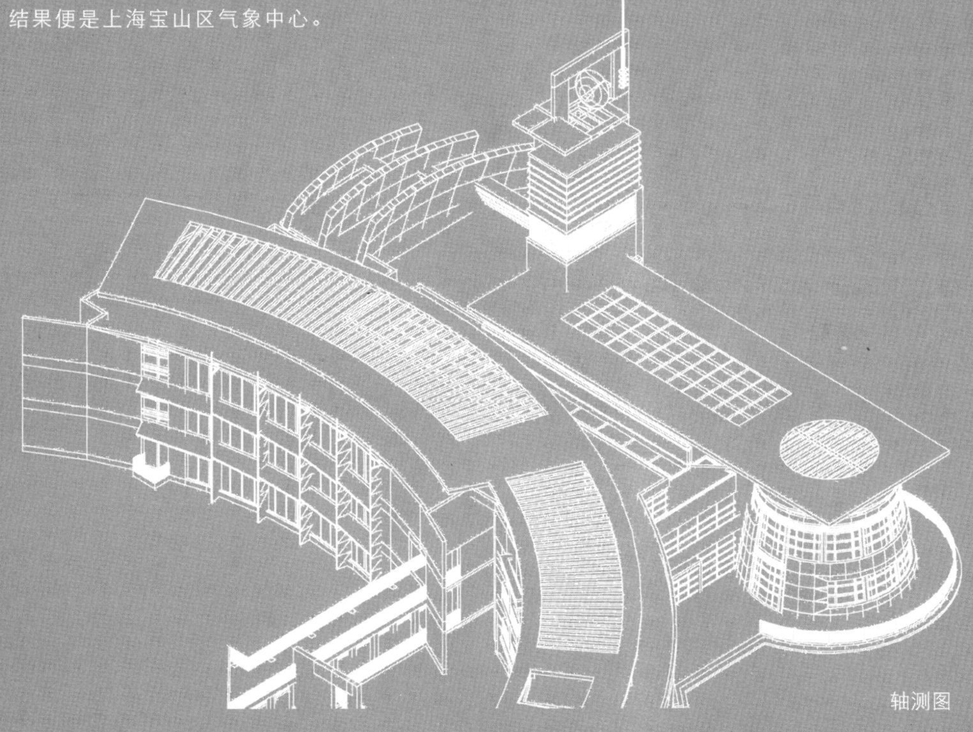

轴测图

第二章

华谏建筑进行时

Chapter 2

DHGP in Progress

Academic Exchange Centre, East China Normal University

上海华东师范大学学术交流中心

Shanghai, China, 2004.

华东师范大学学术交流中心地处闵行新校区东南角，北侧为东川路，西侧为虹梅路。建筑分主楼和附楼两部分，总建筑面积为23,898m²。主楼地上15层，地下1层，建筑高度为53m，属于一类高层公共建筑，功能为四星级宾馆；附楼3层，为国际会议交流中心。

学术交流中心基地位置靠近学校的规划河道，周围环境与景观良好；同时位于城市道路的转角处，附近有规划城市轨道交通经过，交通十分便利，因此在学校总体规划布局上占有重要地位，是学校对外展示的窗口。基地呈东西向长南北向短的结构布局，北侧有学校景观河道经过。依据建筑功能分化，把有大量性人流的各功能体水平方向展开布置，既合理地顺应基地关系，又最大限度地争取到好的景观日照条件。在主楼和国际会议中心之间设置两个功能体的主入口，不仅很好地解决了建筑的内部与外部人流交通的梳理，而且把校园景观引入到建筑内，并透过建筑引入到建筑南向的主题绿地，达到了很好建筑景观效应。

学术交流中心的建筑方案经历了十多轮的磨砺，才最终定稿为现在实施的形态，设计过程十分艰辛。其过程不仅仅是设计师与业主一个思想碰撞的过程，同时也是一个设计自我涅磐的过程。

01 定案设计手稿，倾注了设计者真实的艺术情感。
The final sketch is infused with contains the real artistic emotions of the designer.

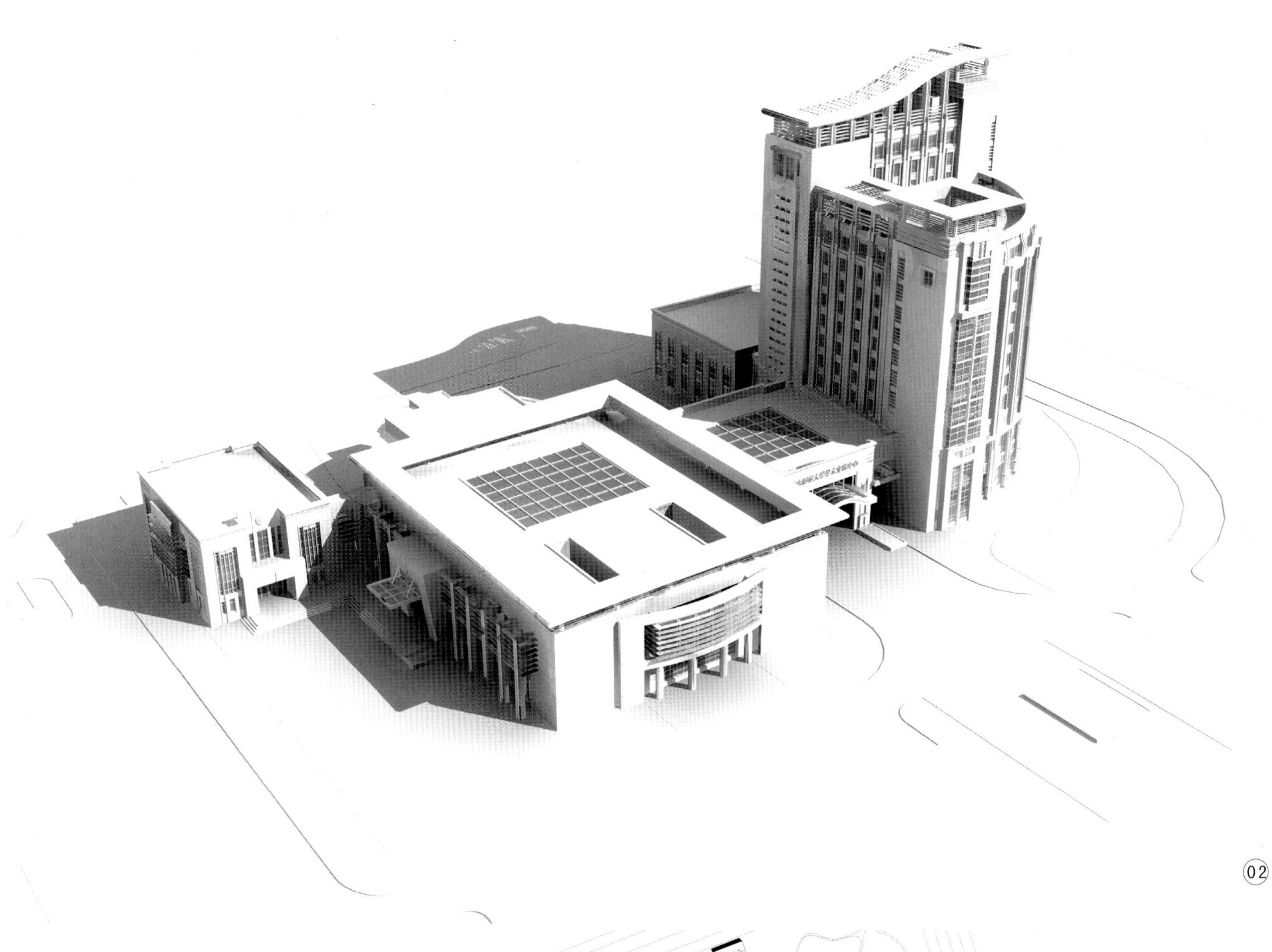

02

Site plan / 总平面

基本资料

地理位置：	上海市闵行区
用地面积：	32,433m²
建筑面积：	23,898m²
占地面积：	5,589m²
建筑密度：	17.23%
容积率：	0.656
绿化率：	35.02%
设计时间：	2004 年

BASIC INFORMATION

Location:	Minghang District, Shanghai
Base area:	32,433 m²
Building area:	23,898 m²
Site area:	5,589 m²
Building density:	17.23%
Plot ratio:	0.656
Green ratio:	35.02%
Time:	2004

02 鸟瞰三维模型，反映建筑本体及周边环境的总体关系，将入口设置在交流中心与国际会议中心之间，由城市道路进来，直接面对校园景观河，取得了最好的入口景观效果。

Airscape of 3D model, reflects an entire relations between the building and its surroundings. The entrance hall is planned to sit between the exchange center and the international convention centre. It is just along side the urban road and faced with the river on campus, which makes a good view effect.

03 鸟瞰在城市街角的学术交流中心建筑群。

Airscape of Academic Exchange Centre on the corner of the cross.

ECNU Academic Exchange Center is located in southeast of the new campus in Minhang District, south of Dongchuan Road and east of Hongmei Road. The center consists of a main building and a podium house, with a total area of 23,898 square meters. The 53m high main building with 15 stories on the ground and one story underground, is a first-class public high-rise, serving as a four-star hotel, while the podium house serves as an international conference and exchange center.

Surrounded by a good environment and view sight, ECNU Academic Exchange Center lies on the corner of urban streets and close to a man-made river. There is an urban rail transit going through it. Thus, the center plays an important role in the master plan of ECNU, which is a window open to outside. With a river passing by north, the base is relatively longer in east to west and shorter in south to north. According to functional distribution, we assign various functional spaces with large streams of people in east to west direction, which may take good advantage of the shape, sight and sunlight. We set entrance of the main building and podium house between them so as to lead inner and outer streams. The campus view is extended into the building therefore expand the theme green land in south, which has a good landscaping effect.

The final design came from more than ten rounds of modifications. The designing process is extremely hard, which is not only a conflict of ideas between designers and our clients but also a pursuit of self-perfection.

04 8 层通高的交流中心内庭是国内高校少见的生态共享厅。
The eight-story courtyard inside the exchange centre is an ecological lobby which is rare in Chinese universities.

05 夜景中的学术交流中心，是城市的夜间新坐标。
Night picture of Academic Exchange Centre, a new landmark in the city at night.

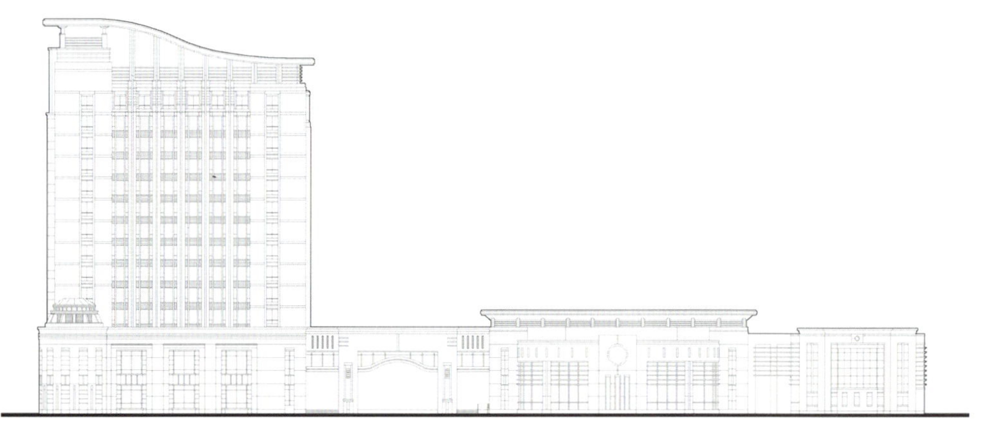

North facade / 北立面

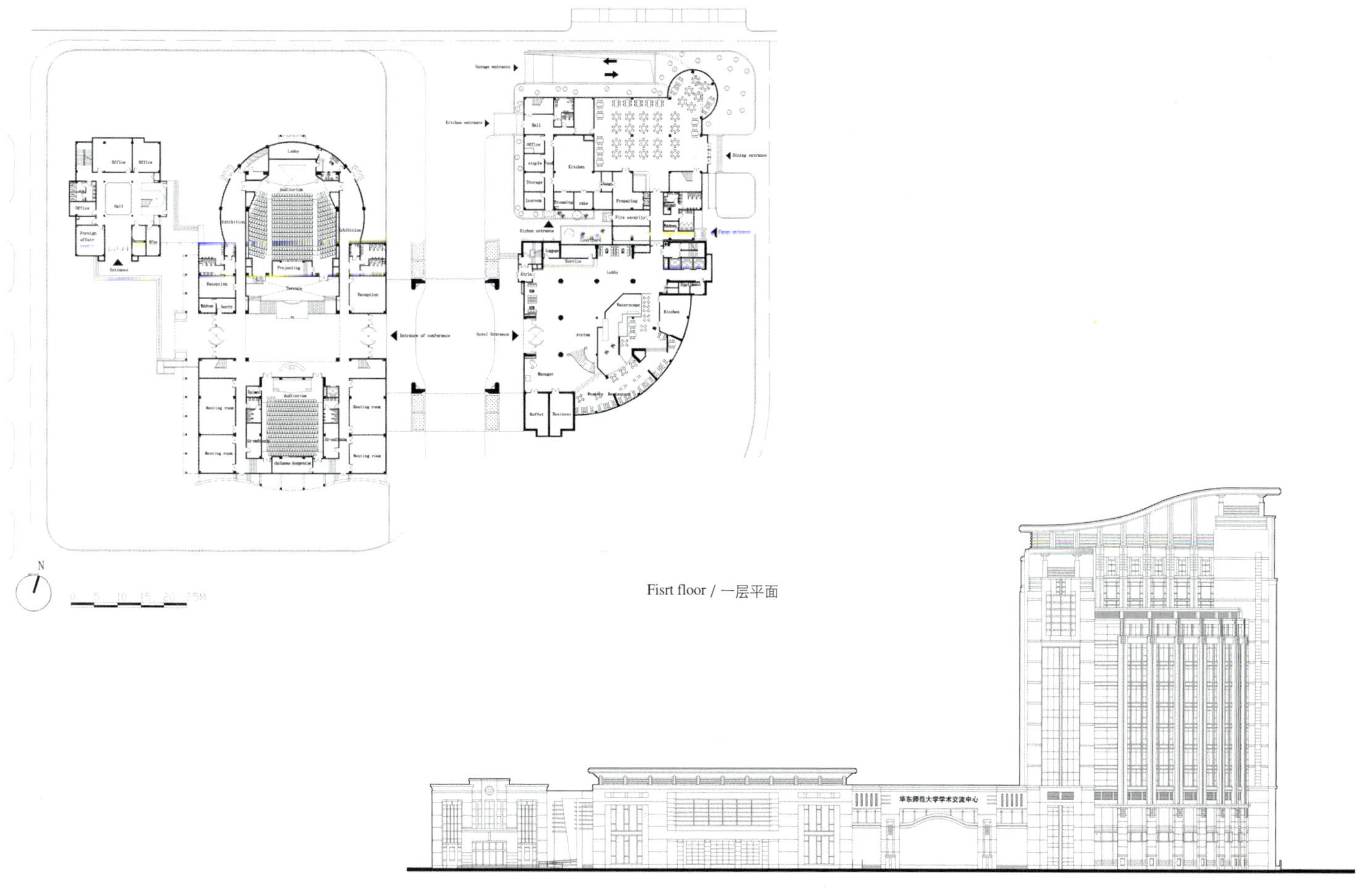

Fisrt floor / 一层平面

South facade / 南立面

漫谈上海“万国建筑”风格化

——华东师范大学学术交流中心设计侧记

首先，我们需要明确一个事实，一个隐藏多年，却不曾溜出嘴边的事实——

当年远东第一银行外滩中国银行大楼、远东第一舞厅百乐门、远东最大的华商证券交易所、亚洲最佳音响上海音乐厅、中国第一游乐城大世界都是中国建筑师设计的。

由来

一个设计作品的酝酿与实现总是一路荆棘的过程，华东师范大学学术交流中心风格选择的过程更是一个经历了从现代到历史的跨越。

“万国建筑”不是什么新鲜的字眼了，或许在那些自以为了解中国历史的人眼中，那是埋藏在心底永远不愿揭开的伤疤，是外国列强蹂躏践踏后留下的耻辱烙印。似乎一提起“万国建筑”，就会唤起中华儿女满腔的爱国热情，不免会有“醉卧沙场君莫笑，古来征战几人还。”的悲壮与感慨。

多少年过去了，这份沉重对于“万国建筑”是多么的残忍，他背负了百倍甚至千倍于自身所能承载的巨大压力。我们为什么不能用欣赏艺术的眼光，怀着一颗包容的心为已经老去的“曾经”减负呢？更何况，“万国建筑”中，一大批主要的代表作品（如本文开头所列）均出自中国建筑师之手！

在当时的社会状态下，老一辈中国建筑师忍辱负重，艰苦创业。黑夜给了他们黑色的眼睛，他们却用来寻找光明！即使再恶劣的环境，也依然无法掩饰他们的才华，一次次的脱颖而出、一次次的光芒四射证明了中国建筑师绝不输给外国建筑师！

我们回顾那段历史，才发现老前辈们呕心沥血，不惜血与泪的无私付出，为的就是给中国现代建筑留下仅有的一丝血脉，他们耗尽了最后的气血以期延续中国建筑的魂。“万国建筑”就好比他们留下的“革命孤儿”，父母为中国建筑的延续付出了毕生的血与泪，难道我们不应去珍视、保护他们的孩子吗？

华东师范大学学术交流中心的设计，众多参与者可谓全情投入，按照过往的经验是一种选择，不加考证，无意识地给她附加上一种大家相认同的方向，但有时候我们发现缺少了对历史价值思考，哪怕是一丁点的回忆。

历史的昭示

我们发现，上海的“万国建筑”，不再是一个个孤立的建筑物，而是一部被时间主线串联起来的历史故事。而无言的历史，则由凝固了的音乐——建筑，向我们娓娓叙说。

我们追踪中国第一代建筑师在上海的足迹，找到了许多鲜为人知的历史事实。一直因误读产生的历史成见，将被修正过来，对“万国建筑”，有了新的认识和评判。上海有一大批著名建筑，是杰出的中国建筑师主持设计的。他们娴熟的手法，完全可以与西方一流学院派建筑师媲美。南京大戏院（上海音乐厅），是西方复古主义建筑风格的一个杰作。1941 年建成的美琪大戏院，注重观众厅的声、光、暖效果，表现出全新的现代建筑风貌。大上海电影院以 8 根霓虹灯柱形成立面，被赞誉为“匠心独具，美妙绝伦”。

而相当多的“万国建筑”，中国建筑师也担当了重要的设计工作，由于任职于外国建筑师事务所，再加上当时社会条件的种种限制，他们的名字没有能够出现在设计者的名单中。但是他们的设计才华，融入并凝结在了上海“万国建筑”之中。这是一段被湮没的历史。

上海近代建筑史不断深入探究，这段历史将渐渐清晰起来。

1932~1937年，100位申请开业的中国建筑师，就有相当一部分来自著名的外国建筑师事务所。堪称杰作的佘山大教堂，有一位重要的设计参与者，就是中国建筑师王信斋。因在建筑工程中的杰出贡献，他获得了当时北洋政府的嘉禾奖。

论及上海的"万国建筑"，建筑作品的数量和质量，中国设计师所表现出的才华与水准和外国人相比，毫不逊色，直逼当时国际先进水准。他们年纪更轻，更具锐气，能够更多地将中西文化融合。在梳理历史的时候，我们有重要的发现。过去似乎疏漏了忽略了这个重大的历史事实，今天应当昭示于世：上海"万国建筑"，中国建筑师功不可没！

当年的"海归"

直到20世纪20年代，中国才有了完整意义上的建筑师。

我国第一代建筑师，大多是留学西方专攻建筑学的"海归派"。20世纪10年代，第一批留学生学成回国，20年代、30年代又有大批留学生陆续回国。他们绝大多数选择了当时中国最大、经济最发达的城市——上海，开创事业。

留学美国康奈尔大学建筑系的吕彦直在1925年，他31岁的时候，与40多位外国建筑师一起参加南京中山陵的设计竞赛，结果他的设计方案一举胜过了多位老资格的洋人，成为中山陵的第一设计师。

20世纪30年代末，在上海10层以上高层建筑中，中国建筑师设计的作品占20%以上。在上海市政府近年公布的优秀近代建筑名录中，中国建筑师的作品也占20%以上。著名的建筑，如外滩中国银行大楼、大陆商场（东海大楼）、大新公司（市百一店），均由中国建筑师设计。特别令人引以为豪的是，上海那美仑美奂、构思精巧的影剧院，80%以上由中国建筑师一手设计，如大上海电影院、美琪大戏院、上海音乐厅等。

岁月流逝，建筑犹在。这一切，足见当时中国建筑师无论在数量上和声誉上都已经和外籍建筑师势均力敌。

中国银行大楼，建成于1937年，楼高17层。解放前，由中国银行发行的一些纸币，就是以中国银行建筑效果图为图案，在图案中，中国银行要比相邻的沙逊大厦高，而且高出许多。

中国建筑师陆谦受原主持设计的中国银行大楼高34层，想打造成为外滩的新地标。后因相邻的沙逊大厦业主作梗，才改为17层。这透露出那个半殖民地的社会状态下，中国建筑师在夹缝中奋斗的艰难。中国银行大楼是艺术装饰主义的摩天楼造型，却配以蓝色琉璃瓦中国传统四角攒尖顶，在外滩建筑群中独树一帜。

作为"万国建筑"之一的上海音乐厅，中外闻名。它是谁设计的？我们可以骄傲地告诉你：是我们中国人设计的。设计者是37岁的中国建筑师范文照和32岁的赵深。当年建成之后，非常轰动，被称为亚洲一流音乐厅。半个世纪以来，一直以最佳音响效果闻名中外。即使在今天为了城市整体规划的需要，必须"动迁"，最后作出的方案，只有"平移"。为什么要这么费劲地去平移？原因就是音乐厅的内部结构太完美了，动不得！你想想，半个多世纪以前，我们中国第一代建筑师便表现出了何等的建筑天才！

中国工匠

说到底，上海的"万国建筑"几乎全是中西合璧的，是多种风格的融合。或者说"万国建筑"中蕴涵着海派情调。看起来是西式建筑，却分明透着东方的气息。一张图纸画出随着专家对来，进入施工阶段后，不可能一成不变，需要调整，需要新的创意。要知道，所有的建筑，都是中国能工巧匠用中国的建筑材料，用中国传统的建造方式造起来的。这个过程是中西风格融合的过程。

几乎所有著名的上海“万国建筑”都是中国技术工人建造。比如外滩的海关大楼，又比如南京路远东第一楼棗24层的国际饭店。

在历史档案中，我们获得了一些珍闻。1932年6月，国际饭店即将进行主体工程。外国营造商预料，这巨大的工程，定会落到他们手里。因为中国建筑施工营造商的设备和技术不如他们。但业主选中的是中国人陶桂林。

那位主持设计的外籍建筑师对中国营造商一向十分信任，但是此番中国人要承建这远东第一楼，却让他产生了几分担心。他认为钢框架的安装，难度很大，于是强调必须等待德国西门子派来的专家抵达上海指导，才能动工。可是德国专家迟迟不来，延迟了地面开工的日期。中国营造商决定自己动手。当西门子专家赶到上海，大吃一惊，钢框架已经安装到了11楼。洋人怀疑质量有问题，然而经过严格检测，全部符合要求。从此洋建筑师们对中国技术工人的工艺水准，刮目相看，钦佩不已。事实上，中国营造商派出的工地主任，都是工程师级别的专业技术人员。他们不是死板地依照图纸施工，而是在施工过程中不断地解决现实难题，从某种意义上是担当了后期设计的完善工作。

自从1883年川沙人杨斯盛开设的杨瑞泰营造厂成功地承建海关大楼之后，名声大振，中国人开设的营造厂大批出现。从20世纪20年代开始，中国人的施工专业队伍垄断了上海的建筑施工市场。一幢幢让人惊叹的“万国建筑”就是在这些优秀的中国建设者的“智慧和艰辛的创造中”拔地而起。

华东师范大学的自豪

走在夕阳暖照里的上海大街，看着那些染着岁月沧桑而不减端庄巍峨的老建筑，眼前浮现出那些卓有成就的第一代建筑师的年轻面容。在泛黄的照片上他们有的西装革履，有的身着长袍。面带倦容的他们丝毫不显年轻人的张狂与热情，可是在他们的作品中，全然没有过分老成的影子。

欧阳修的《蝶恋花》有“泪眼问花花不语，乱红飞过秋千去”的妙句。霎那间的回眸一掠或许比得上永生永世的全神贯注，不经意的自然流露好过千重万叠的渲染修饰。冷酷的背后是火焰的激情，沉重的脚下是轻快的华尔兹。建筑成了建筑师谱写的华丽乐章，踩着节拍我们翩翩起舞，任凭时光流逝，我们乐在其中。

穿越悠悠岁月，我们的心底不禁涌起深深的追念和敬佩之情，禁不住用手去轻轻触摸那些老建筑沉默、厚实的花岗石墙面，似乎触摸到了历史。

言语间我们已经走过上海建筑的百年，上海“万国建筑” 风格化在我们眼里也许只是一个被我们用来记录的符号，但在我们在华东师范大学学术交流中心的设计过程中我们一路走来，正因为历史的作用，发现我们现代的设计师不再贫乏，我们有东方神韵，我们有宽广胸怀，我们有时候甚至无所畏惧，因为在华东师范大学这片土地上我们正朝着一个上海“万国建筑”的新地标努力跋涉。

设计过程

由于华师大学术交流中心的多样和复杂性，给设计带来了较大的困难和挑战。华东师范大学闵行新校区学术交流中心从方案设计到施工图设计，共经历了十轮的设计过程，历时一年半。

回顾整个设计过程，可分为三大阶段：第一阶段（第一至第二轮方案），设计理念为“集中式”的平面布局和现代主义，甚至是高技派的现代主义立面造型；第二阶段（第三至第七轮方案），设计理念上发生了重大变化，即平面布局由“集中式”转变为“分散式”；立面造型由“现代主义”转变为“现代典雅主义”。第三阶段（第八至第十轮方案），设计理念上虽然传承了第二阶段的设计思路，但设计手法和元素上，则更加简洁、明快和开放。

对两种设计理念的切身体会

在华师大学术交流中心的设计过程中，我们始终把功能的合理性作为方案构思的出发点和归宿。第一阶段的设计中，我们很好地解决了“尽量减少交通面积”的问题。但是，由于该项目功能上的多样和复杂性，带来了许多问题。例如，不同功能分区之间的界线不太明朗，诸多人流之间的交叉和相互干扰，疏散线路过于集中甚至冲突等。这些问题在某些方面制约了方案的深入。所以，第二和第三阶段，大胆地放弃了“集中式”的设计理念，改为“分散式”的设计方案。把学术会议、国际交流和接待中心三大功能区明确地划分成三大块体，相对独立。相互间用连廊相通，联系又方便；尽管这样交通面积稍许多了些，但解决了很多问题。首先，人们从单体外观特征上，可直接判断哪个是会议中心，哪个是接待中心和国际交流中心，进而直接到达自己的目的地，避免了“集中式”建筑的“先进入，后四处寻找”的盲目性；其次，由于我们把功能复杂的学术交流中心分隔成功能相对单一的三个独立体。因此，人们到达某一单体（目的地）后，可清晰地看到内部的结构布局和功能关系。这对于避免人流的交叉或走回头路，以及紧急疏散时的“畅通”方面非常重要。再次，因为每个功能区自成一体，相对独立。所以，日常的管理上也可单独管理，独立运作；并且对于设备运行的控制和节能上也都有利… …

华东师范大学学术交流中心地处两条城市干道交汇处。所以，作为国际和国内交流而设置的公共建筑，交流中心的效益是建立在与城市空间相互共融的基础上的。因此，设计中创造积极的公共空间，如，集散广场等。把城市空间引入建筑基地内，提供良好的停车及步行空间，保持交流中心与社会大众之间的便捷联系。基于这种设计理念，把“接待中心”（准四星级宾馆）设置在距两条道路均相对近的一侧，把“国际会议中心”设置在基地的中央，突出 “学术会议中心”的项目特性。“国际交流中心”主要服务对象是到华东师范大学留学生，以学校内部联系和交流为主，因此设置在离校区最近的一侧 。

各建筑单体体形上，我们之所以从现代风格的建筑设计转变为现代典雅主义的建筑设计，主要目的是避免纯现代建筑因片面强调功能，无视历史、文化及建筑内涵，致使建筑形式单调乏味，缺少人情味；这也体现了华谏国际“后历史主义”的一贯设计思路；在充分尊重功能合理的基础上，广泛吸取现代建筑的一些理念、思路和手法，丰富建筑形式，突出建筑文化及内涵。

外装修处理也是关系到外观效果的一个至关重要的方面。设计遵循典雅主义搁 问綌造型，分别采用石材，仿石喷涂和铝塑板等，充分演绎三段式的设计特征。在强调体现各建筑单体性质特征的造型基础上，重视细部的处理。如，引入各种线角、细分外墙及柱子的分格等，来加强建筑整体形式与细部的比例关系，使本学术交流中心既典雅大方，又极富魅力。

The campus updating program of USST

上海理工大学校区更新计划

Shanghai, China, 2004.

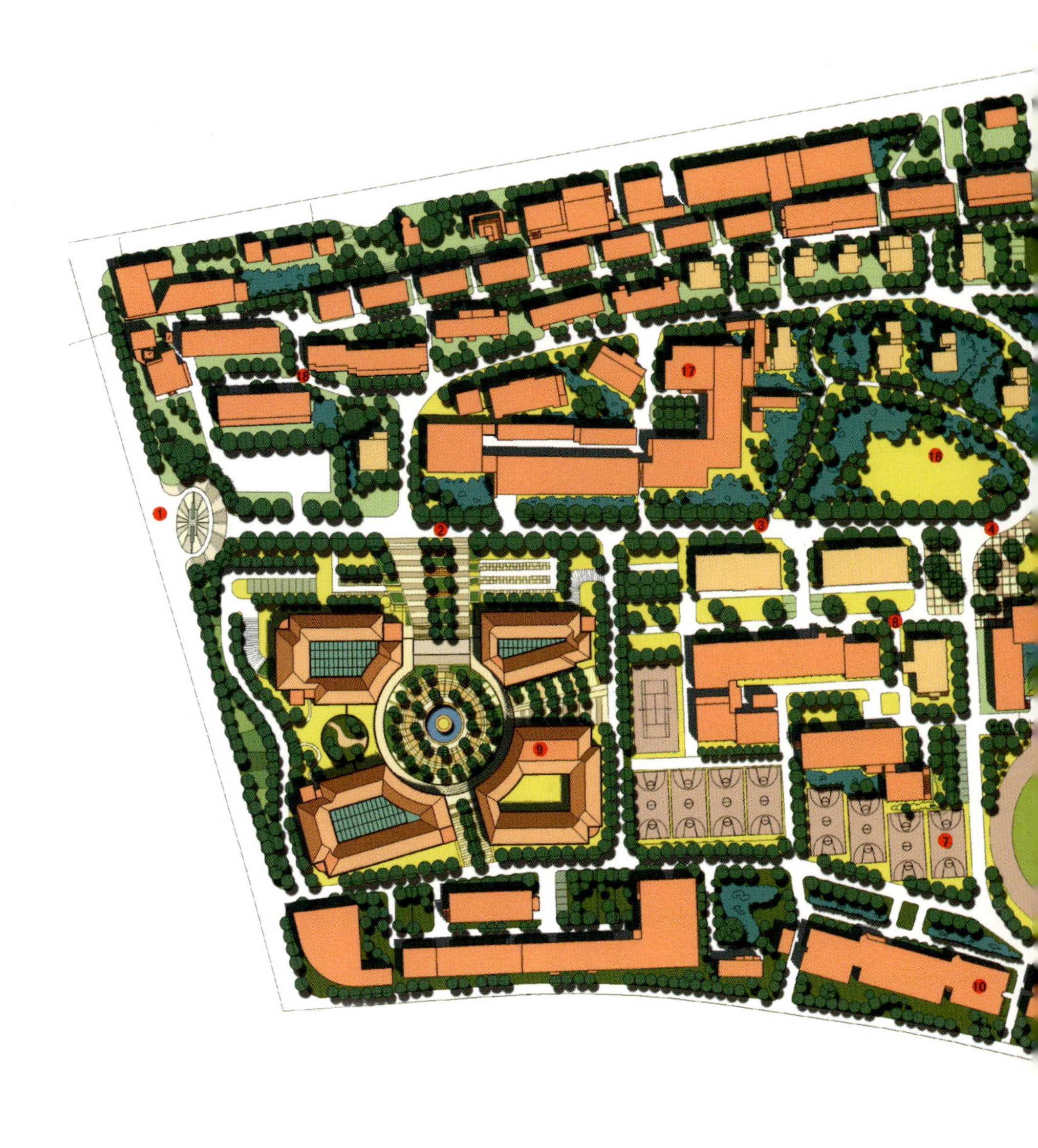

上海理工大学（原沪江大学）创建已有百年历史，是以工科为主、兼有文、理、管、经、医的多科性综合大学。校本部位于杨浦区军工路516号，东临黄浦江(与复兴岛隔内河相邻)，西靠军工路(建设中的中环线)，南接海安路，北至虬江河。学校占地面积约37万m^2。

上海理工大学原沪江大学校址地段保护更新计划 作为具有百年历史的上海理工大学，其前身是1900年由美国教会组织侵信会所创办的上海侵信大学，是一所地道的教会学校，后逐步发展为一所综合性大学，1915年定名为沪江大学，1929年纳入国民政府教育部。校区内1949年以前建成了大批哥特式建筑，是中国哥特式建筑保存最完整的大学校园，有着丰厚的历史底蕴以及优良的学术传统。步入20世纪90年代，上海市逐步将校园内的大批优秀历史建筑纳入上海市历史保护建筑行列。对原沪江大学校址地段的哥特建筑群的保护更新并不完全是抽象的，它需要体现在每个师生的生活方式之中，体现在校园的整体环境之中，体现在历史与现代的对话中，表达了对整个校园的历史文化和空间环境的深刻理解和感悟。

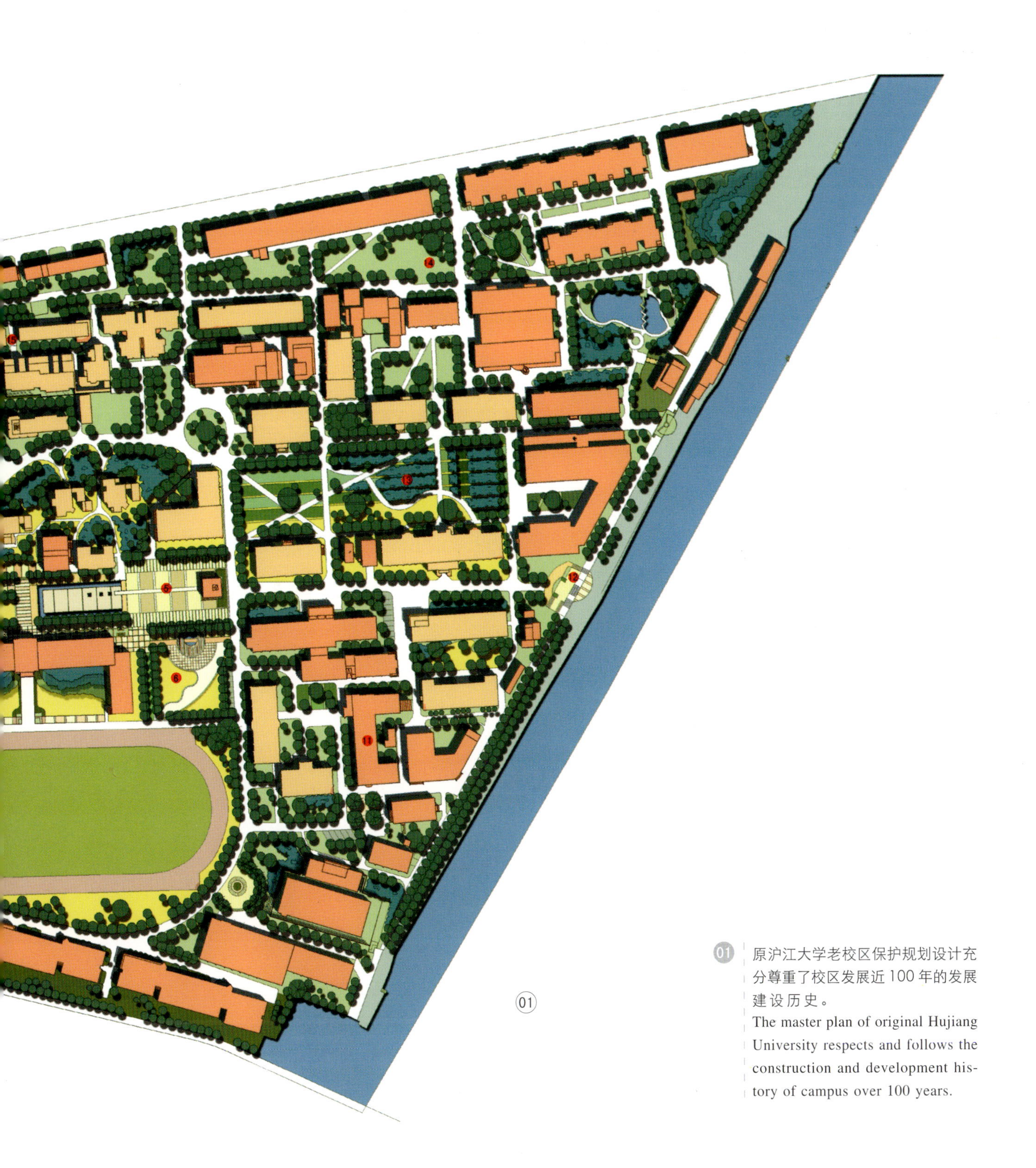

01 原沪江大学老校区保护规划设计充分尊重了校区发展近100年的发展建设历史。
The master plan of original Hujiang University respects and follows the construction and development history of campus over 100 years.

The University of Shanghai for Science and Technology (original Hujiang University) has a history of over 100 years. It is a full-fledged multidisciplinary university committed to training of professionally competent personnel in the fields of engineering (its main feature), management, commerce, arts, science and medicine. The main campus is located at 516 Jungong Road, Yangpu District, which is adjacent to Huangpu River to the east (next to the inner river of Fuxing Island), Jungong Road to the west (the under constructing middle circle line), Haian Road to the south and Qiujiang River to the north. The whole campus covers an area of 55,100 square meters.

The campus upgrading and protecting program of USST (original Hujiang University)

With a history of over 100 years, USST was formally an ecclesiastic school founded in 1900 and then grew up to be a multidisciplinary university. In 1915, it was named after Hujiang University. Education Ministry of Chinese government took over it in 1929. A great number of Gothic buildings were built on campus before 1949, which made it the best-reserved Gothic buildings of all Chinese campus. In 1990s, the local government started to list the old buildings in USST as local historic protecting buildings. It is not wholly abstract to protect and renew the Gothic buildings in USST. It also needs to emerge from staff and students' life, from the whole environment on campus and from the conversation between the old and the new. The project tries to express a deep consideration and thinking to the historic culture and spatial environment of the whole university.

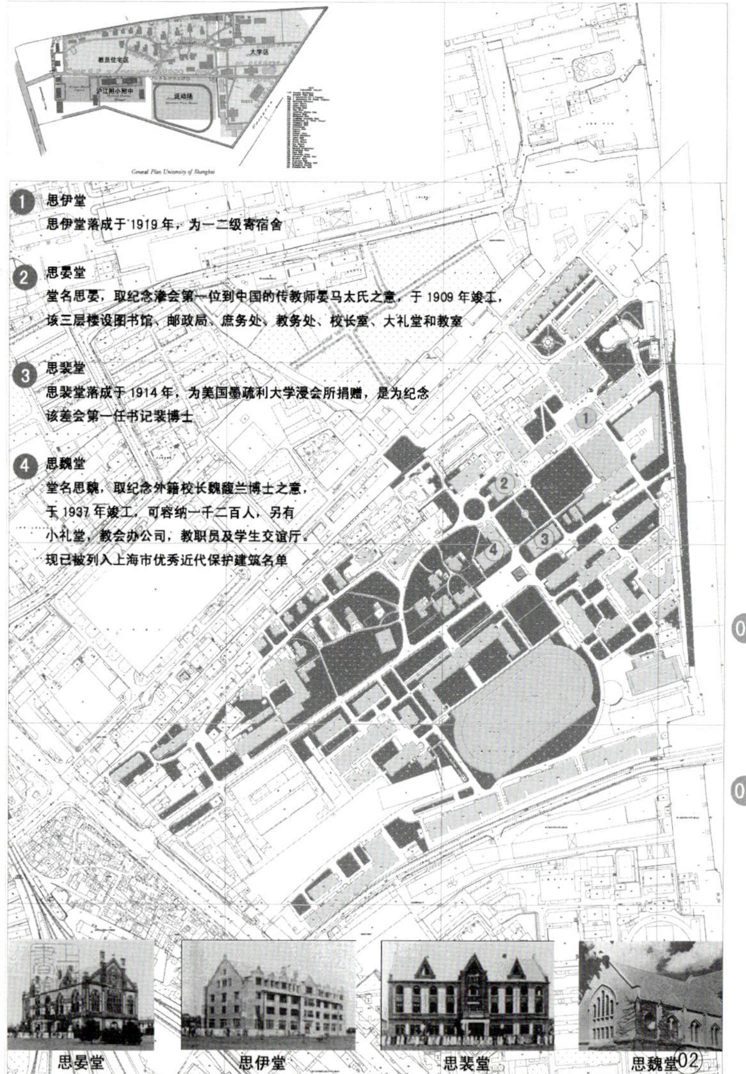

02 对校区形成初始阶段的研究，是整个老校区保护规划的重要依据。

The study on initial stage of USST is an important basis of reservation and plan of old campus.

03 将校区发展史所经历的历史阶段的建筑符号系统整合到一个较完整老校区核心景观区中，对不可逆的历史进行可逆的保护。

The architectural signs that were formed in the history of USST are integrated into a key landscape area in old campus. We try to reversibly protect inevitable history.

新建教学综合楼 位于老校区主轴右侧紧临校园主入口，建筑被设计成一个建筑内部和外部空间相互联系的综合体。建筑由四栋教学楼组成，以一个内部独立圆形连廊串联成一体，南北、东西向两条轴线贯穿全局。东西向轴线延续了原有老校区的轴线，继承和发展了整个校区的空间形态；南北向轴线平行于军工路，与城市肌理相适应，形成良好的沿街景观。建筑内部的圆形广场是一个戏剧化的设计，三层高的圆形走廊围合出可以进行交流、对话、观演等多种活动的场所，成为建筑群的高潮。

教学楼建筑造型延续老校区的文脉，采用古典三段式构图，体量设计对称而富有变化。立面运用古典主义线脚、壁柱等种种构图元素，丰富而精致。厚重的墙面与细腻的线脚相互呼应，精准推敲的细部耐人寻味，处处体现了教学楼庄重严谨而不失活跃气氛的性格。

新建图文信息中心 位于新老校区的交接处，是校园发展轴的收尾建筑。新建的图文信息中心在报告厅的上部设置了一个大型的半公共性的共享中庭，作为图文信息中心这个建筑综合体的入口，起到了有效组织交通、促进交流，休息以及展示的作用。共享中庭对外开敞，采用玻璃透光顶棚，布置室内绿化。中庭是图书馆这一特定场所的中心，它呈现出三维视觉感和四唯的意境感，其具有的通透性使中庭与内院、中庭与外部广场形成连续的视线走廊，从而构成一个有机的整体。

New teaching complex building

The new teaching complex building lies on the right side of the principal axis of the old campus, next to the main entrance of the school. The building is designed as a complex, where the inner and outer space are well connected. The complex consists of four parts with an unattached round corridor inside connecting each other. The south-north axis and east-west axis run through the whole complex. East-west axis succeeds to the old axis of the campus, inheriting and developing the space modality of the campus. South-north axis parallels to Jungong Road, which has a good match with the street-scenery of the city and a good combination with the city skin. The round square inside is considered as a theatrical design and the highlight of the complex. Surrounded by 3-level round corridors, the square is planned to provide a space for conversation, communication and performance.

The style of the new teaching complex building is a continuation of the old campus. With the classical three-composition style, the building looks various and symmetrical. With the classical lines and pillars, the facade looks magnificent and delicate. The thick walls, delicate lines and precise details embody sublimity, preciseness and activity of the teaching building.

New Information Center

Located between the old and new campus, the new information center is the last building of the developing axis on campus. A large semi-public courtyard is in the upper floor of the central report hall. As the entrance of the information center, the courtyard is planned to organize traffic, promote communication, relaxation and exhibition. The courtyard has a grass ceiling and is decorated with in-door plants. It is the center of the library, which provides a 3D visual sense and a 4D artistic conception. The continuity of the courtyard turns inner space and outer space into continuous visual corridor. Thus, form an organism.

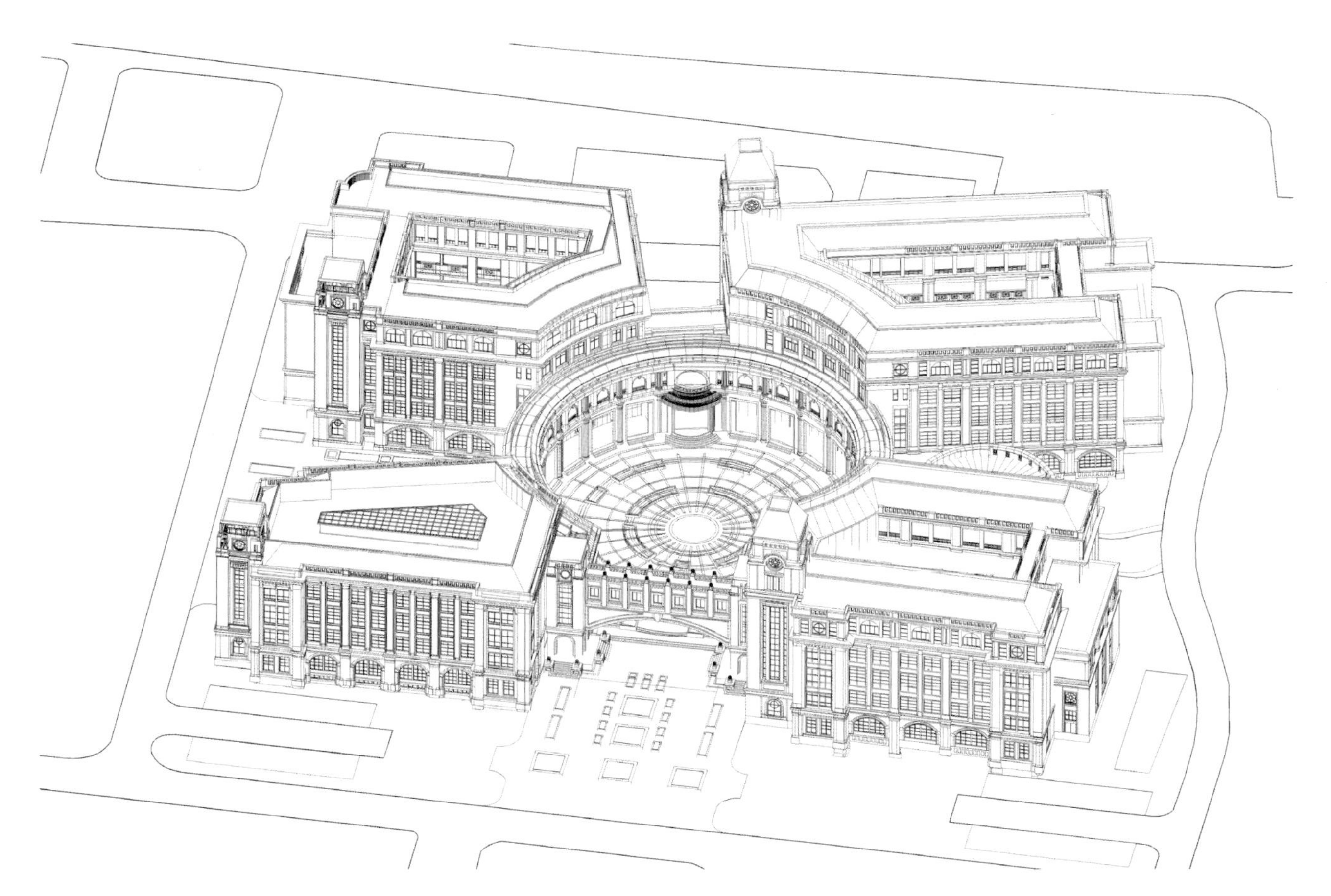

3D model/ 教学楼 三维模型

05 教学楼临着城市干道一侧的建筑立面层次丰富、形象古朴凝炼。
Decorated by various layers, the facade facing the urban road looks classical and dignified.

06 抽象古典建筑的柱式。
The columns are abstracted from classical buildings.

07 教学单元角部的塔楼。
The tower on the corner of the teaching complex.

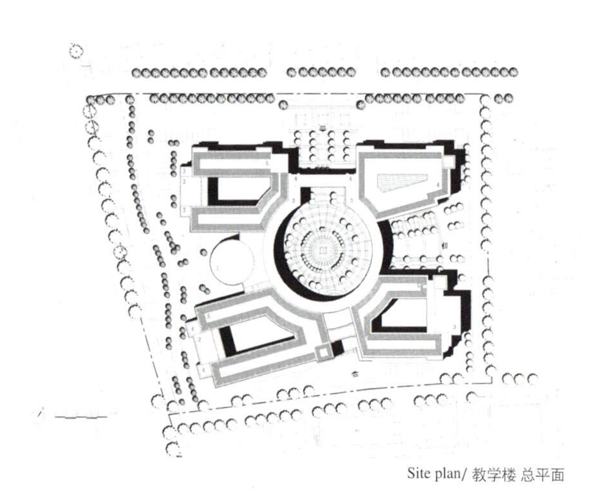

Site plan/ 教学楼 总平面

老校保护区基本资料		Basic data of old campus:	
地理位置：	上海市杨浦区	Location:	Shanghai Yangpu district
总用地面积：	约300亩	Gross base area:	300 acre
新建教学综合楼基本资料：		New Teaching Complex Building	
用地面积：	23,260 m²	Base area:	23,260 m²
建筑面积：	30,929m²	Building area:	30,929 m²
占地面积：	9,050m²	Site area:	9,050 m²
容积率：	1.33	Plot ratio:	1.33
绿化率：	37%	Green ratio:	37%
设计时间：	2004年	Time:	2004

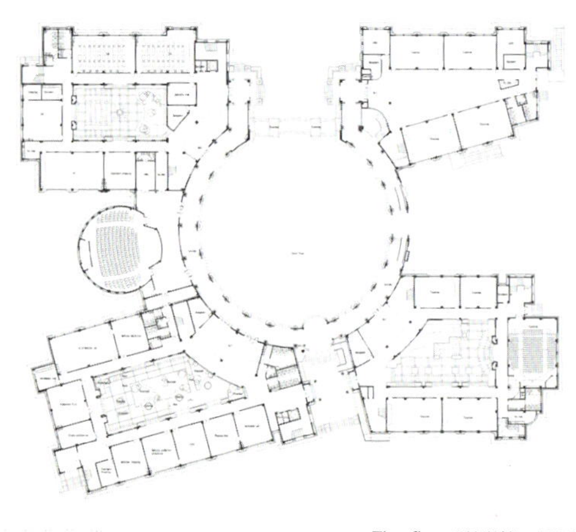

Fisrt floor / 教学楼 一层平面

06

08 设计师注重艺术化表达的手绘稿。
Designer's brush work, emphasizes artistic expression.

06

以空间传承历史

——上海理工大学新教学大楼与图文信息中心设计

回 顾

“对于1926年来访问中国的人而言，如果旅行者是从上海进入这个国家，在城郊外他的轮船就会经过一所学校的建筑群，他会被告知这是由美国侵信会办的沪江大学，大学校园四周都栽着柳树和四季长青的灌木，校场的中央，巍巍地布着二三十幢洋楼。在茶余饭后的时光，同学们更喜欢三五成群，或则一个人，在黄浦江边散散步，在草地上小坐一刻，看看天上的夕阳和白云，听听江里的潮声，俨然是一幅从弗吉尼亚移来的世外桃源景象。”

这些文字就是《沪江大学的历史》一书中关于沪江大学——上海理工大学的前身，昔日宁静江畔校园的描述。

上海理工大学由来已久，文脉源远流长。早在1906年，由美国南北侵信会(Northern and Southern Baptists of America)创建，初名上海侵信大学(Shanghai Baptist College)，是地道的教会学校，后逐渐成为一所综合大学。1915年改名为沪江大学(University of Shanghai)，并于1917年在美国弗吉尼亚注册立案。此后历经战争、迁徙、调整，至1952年，当年的沪江大学校园幸运地被完整保留，并移归后来的上海机械学院，也就是现在的上海理工大学。和许多上海其他的老房子一样，如今这片校园也都被挂上了保护建筑的铭牌，以纪念那昔日的“光辉岁月”。

缘 起

时至今日，在举国上下高等教育迅速发展的大背景下，上海理工大学新教学大楼与图文信息中心的建设被提上议事日程，校园在原有基础之上往北拓展作为图文信息中心建设用地。

对于具有如此深厚历史背景的一所大学，建筑师如何寻求新建筑在百年校园之中的定位？

在传承与颠覆之间，我们选择了前者，因为历史的博大精深，和她历久弥新的巨大能量！

解 读

理 念

在不断的创作中，我们发现能够永久留传的，是某种精神上的东西，而那些散发着历史厚重感的建筑，是唯一适合承载这种精神的空间载体。当岁月流逝，物是人非，爬满红色清水砖墙上的斑驳青苔，和圆形柱廊中的冬日光影，将向无数后来者诉说她不曾断裂的辉煌历史！

建筑作为一个可以表意的个体，是通过表皮与空间来传达讯息的。而建成环境的最终意义，是实现人与建筑，建筑与周边环境在时间维度上的和谐共生。因此，我们关注的不仅仅是建筑立面，更有室内外空间、建筑中人的活动以及它们在过去、现在、将来随时间流走而留下的印迹。上海理工大学新教学大楼与图文信息中心的设计，便是基于这种考虑所做的一种深思熟虑的选择。

表 皮

上海理工大学校园内现存1948年前建造的各类建筑物30余幢，建筑面积约30000m^2，是目前上海保存最完好的教会建筑。校内钢窗的使用，首开上海近代建筑的先例。这些建筑多清水红砖砌筑，两坡红瓦屋面，风格统一，砖混结构，二至四层不等，是学校教学科研用房和教职员工住宅、学生宿舍等。其中，仪表楼、大礼堂早在1994年由政府公布为“上海市优秀近代建筑”。

国外大多数著名大学的校园建筑在艺术上的一个共同点是对学校传统的重视。校园建筑首先是属于某个特定的校园环境，这一点已成为人们的共识，而任何一所真正好的大学都是建立在浓厚的传统根基之上的，这一点也为人们所公认。各名校都把维护校园的统一面貌、保持学校建筑的传统风格看成是非常重要的事情。

在上海理工大学这两栋新建筑的设计当中，对表皮的处理更多地考虑到能否与周边建筑，特别是历史建筑实现对话，浅黄色石材与红色砖墙成为主导材质，他们丰富的质感与钢和玻璃的光滑透明产生强烈的对比，在材质、色彩上我们极力突出建筑的厚重感、历史感与文化感。永恒的三段式立面，经典而恰到好处的细节装饰，令人真正体会到帕拉第奥所说“美产生于形式，产生于整体和各个部分之间的协调”。当表皮这种形而下的建筑元素直接作用于人的视觉，人们会自然地将其关联到那些封存已久的历史，并追问其历经沧桑的点点印迹。

空间

20世纪30年代沪江大学的校园空间布局大抵如此：从军工路校门入内，一条林荫道直通复兴岛运河，林荫道南侧依次是附小，附中(思乔堂、思孟堂和思雷堂)，再往里为运动场，北部为排列有致的小洋房建筑群，为教员住宅区。东端为大学区，原有教学楼一幢（建于1908年的思晏堂），学生宿舍两幢(思裴堂、思伊堂)，女生宿舍两幢，体育馆、健身房、科学馆、图书馆、大礼堂（思魏堂）各一幢。所有这些建筑及其所形成的校园空间到现在为止，与当时相比几无变化，都被幸运地保存了下来。所不同的是，经过近一个世纪的时间雕琢，她的内在已经越来越丰盛而饱满。

新建筑的兴建，对于这个百年校园整体空间的发展完善是一个很好的契机，于是我们多了一个整合校园空间、梳理校园历史的机会。

建筑空间有内外之分，建筑内部空间我们强调功能合理、使用舒适，这其中我们更关注建筑对人的关怀。通过中庭、门厅、回廊等空间的塑造来满足不同时间、不同人群的活动。注重光线，特别是阳光，从不同角度、不同方位的引入。配合室内绿化、休息座椅等人性化设施让人有温暖舒适的空间体验。

外部空间则以整合校园空间为目标，通过对校园整体空间的梳理，我们明确了林荫道的

教学楼效果图

空间主轴地位，而图文信息中心则是新老校区联系轴线上的标志性建筑，新教学大楼作为老校区最大体量的建筑物在处理其与校园空间的关系上颇具难度，为了减小教学大楼的空间尺度，我们将其分散为四个带中庭的教学、实验单元，既满足了分期建设的需要，又围合形成极具震撼力的圆形广场，通过面对林荫道的入口广场以及面对军工路的开口，使教学大楼内部的功能性开放空间与校园主轴、军工路沿街绿化产生联系，它完美地实现了该地块空间与校园整体空间乃至城市空间的呼应与对话。对于图文信息中心，由于地处校园拓展用地的中心，与老校区中心广场遥相呼应，一种大体量的集中式布局就此产生，它是老校区通往新校区的空间收束，是一种集聚的能量，向四周发散以突出其标志性。

在所有与建筑相关的空间当中，最感动人的，是处于内外之间的灰空间，因为它介于人工与自然、安全与危险、熟悉与陌生之间，可以同时满足不同人的不同需求，在天空、树木等自然背景的渲染下而更富感染力。新教学大楼中拱形天桥对入口的限定、圆形拱廊内的交往空间及其围合而成的圆形广场；图文信息中心中的入口空间、类似哥特式建筑飞扶壁的侧柱所形成的柱廊空间、报告厅外围的扇形柱廊等，所有这些由古典元素塑造的空间都在试图表达一种永恒的时空概念：它承接过去，开启未来。它会在时间中渐渐风化旧去，但也正是时间，将赋予它持久的生命力！

活动

早年的沪江大学无疑是昔日上海小资生活方式的摇篮，大学生们多为一帮十分懂得生活的人。作为一家由外国人开办的教会学校，进入这样的大学念书需要高昂的学费，所以沪江大学也就成了当时上海有名的“贵族学校”，学生英文水平普遍较高，学生社团名目繁多，校园生活亦显丰富多彩。

创造一个优雅、宁静、而有浓厚文化气息的学习、交流环境，是大学治学的根本。随着现代教育的发展，学校开始从内向封闭型逐渐往开放型、智能型过渡。教学楼与图书馆作为提供师生文化交流、信息沟通、知识融汇的最主要场所，需要在满足学生与老师交往活动方面做充分考虑。在空间塑造的时候，我们以休息、停留、交谈、讲授等活动为目标，创造不同功用、不同形态与感观体验的空间，以便各种信息可以在不同场所以正式或非正式的方式自由传递。

在建筑设计过程中考虑人的活动，为的是使建筑更适合特定人群的使用。当人的活动与建筑空间以某种完美的方式结合的时候，空间的意义便得到了升华。

缘继

在古罗马人的信仰当中，每一个独立的本体都有自己的灵魂，这种灵魂赋予人和场所生命，同时，也决定他们的特性与本质。空间的灵魂，也就是场所精神。

时至今日，当我们的教育被西方模式化的时候，当我们的城市被现代建筑所包围的时候，我们那来自灵魂深处的某种理想，是否与当年这所教会大学的创办者有所呼应？或许，这也是另外一种意义上的传承？

教学楼室内透视

Shanghai University of Electric Power

上海电力学院平凉路校区

Shanghai, China, 2005.

上海电力学院扩建项目位于上海市杨浦区138街坊。基地呈长方形，范围包括长阳路以南，河涧路以北，东接上海市东区污水处理厂，西面为上海市25中学和居民住宅区，总占地面积约70万 m^2，毗邻学院平凉路老校区。共要建构诸如电力特色专业教育、研究生教育基地；成人教育、国家电力南方高级培训基地及电力科技产学研合作创新孵化基地等扩建项目。

扩建校区项目用地与老校区隔街相望，“8”字形相连，连接部分比较薄弱，因此新校区格局自成体系，且与老校区单线相连。今天的校园正由原来内向的，封闭的，游离于城市之外的隐士型学院，转变为功能复杂、多层次、多元化的社会综合体。在比较独立的小块用地中塑造校园独特的氛围，并形成校园综合群的形象，是电力学院新校区规划建筑设计方案中的基准点。

上海电力学院经过了一个相当长的历史时期而逐渐形成现在的面貌，不同风格、不同体量、不同材料和结构的建筑交错拼贴在一起。老校区的空间形态虽然变化丰富，但是略显杂乱。本次建筑规划设计凭借校园扩建的契机，在老校区紧凑且富于人性的空间特色和已经形成的文化底蕴的基础之上，加以改造、整合，形成收放有致的建筑空间以及前后延续的文化脉络，对整个校园的环境优化与文化继承产生巨大的推动作用。

01 校区的总体鸟瞰图，反映出校区在市中心区所表达出的特定的内向与外向的双重性格。
Airscape of campus reflects a particular introversive and extroversive double personalize of campus that is located in the urban centre.

The expansion project of Shanghai University of Electric Power is located at NO.138 block, Yangpu District in Shanghai. Near the old campus, the oblong base covers an area of 700,000 square meters and lies north of Changyang Road, south of Hejian Road, west of Shanghai Eastern Sewage Factory and east of Shanghai NO.25 high school and residential areas. The expansion area will establish not only a base of professional education, graduate education, adult education and a base of advanced training of State Power in the South but also a production—study—research cooperation base of power science and technology.

The expansion area is one street away from the old campus with a weak-linking area like the number 8. Therefore, the new campus has its own system with a single-line link with the old one. Today, SUEP is changing itself from an inward, closed and isolated university to a multifunctional, multilevel and multi-oriented social complex. To create unique campus atmosphere and image of integrated group is the benchmark of the SUEP new campus' constructive design.

It took quite a long time of Shanghai University of Electric Power to create today's image with a mix of different styles, sizes, materials and architectures. The forms of the old campus is various but disorderly. With the opportunity of the extension, based on the compact space characteristic which is full of humanity and the cultural detail of the old campus, the constructive design will reform and integrate the old campus to form a lasting cultural thread. That will greatly promote the optimization of the whole campus environment and cultural heritage.

A 研究中心与科技楼
B 南方培训中心
C 图文信息中心
D 研究生、留学生楼
E 教学综合楼

A Power Research Centre and Technology Tower.
B South Training Centre of State Dower
C Information Centre
D Students'Dormitory.
E Teaching Complex.

建筑单体解析

图文信息中心

新建的图文信息中心是新校区富有特色的标志性建筑物。建筑主楼为双塔楼，楼高12层，为校园制高点。主楼正对校门，折形环抱中心圆形广场，引导整个校区的序列，体量简洁，造型丰富，成为进入校园的对景核心建筑。

教学综合楼

教学综合楼主入口面向东面设在南北轴线上，南部报告厅偏向西侧结合图文信息中心围合成新校区的南部入口广场空间，这些变化丰富的组群方式为营造良好的校园学习氛围创造了条件。教学楼综合楼围合成院落式，为师生之间的课外交流提供了一个安静的场所，加强了空间的层次性，成为从外部进入教室的过渡。这些聚集人气的非正式交往场所吸引了众多学者学子在此联欢、休闲、激活创新思维、交流科研成果。建筑造型简洁、现代，充满了时代气息。

电力科技楼

电气科技楼由电力系统专业实验室、电力科学技术实验室、电机基础实验室等部分组成，位于基地东北部。平面上与电力研究中心形成咬合关系，建筑形体相互穿插呼应，形成整体。布局上讲究构图的艺术性和形式美，体现各种功能的合理组合，强调布局紧凑，张弛有致，富有节奏感和韵律感，简洁大方而又变化丰富。

国家电力南方培训中心

国家电力南方培训中心位于基地的西北角，紧邻长阳路，并在长阳路上设置了机动车主要入口。中心塔楼依照此区域空间肌理及新校区规划轴线形成两个方向的转折形体，既传承了整个校园的空间序列，又取得了与长阳路的协调关系，形成了良好的沿街造型。主要客房布置为南北向。裙房与长阳路平行，整体造型简洁、现代，与其他建筑互相呼应。

电力研究中心

电力研究中心位于电力科技楼的西侧，在长阳路和校园内部均设有入口。其造型为折线形，与圆形广场呼应，为老校区进入新校区方向的对景。电力研究中心立面纯净，结合一些互相穿插的梁板，造型简洁大方，视觉冲击力强。

研究生、留学生楼

研究生、留学生楼为塔式高层，共15层。平面布局紧凑，体型错落有致、简洁大方，体现了现代大学生活的时代气息。

一个次生的新区圆形核心广场，一条指向老校区主轴线，表达出新建校区与老校区功能上的连续与更新。

A new round central square and an avenue leading to old campus, indicate a continuation and renewal of old campus.

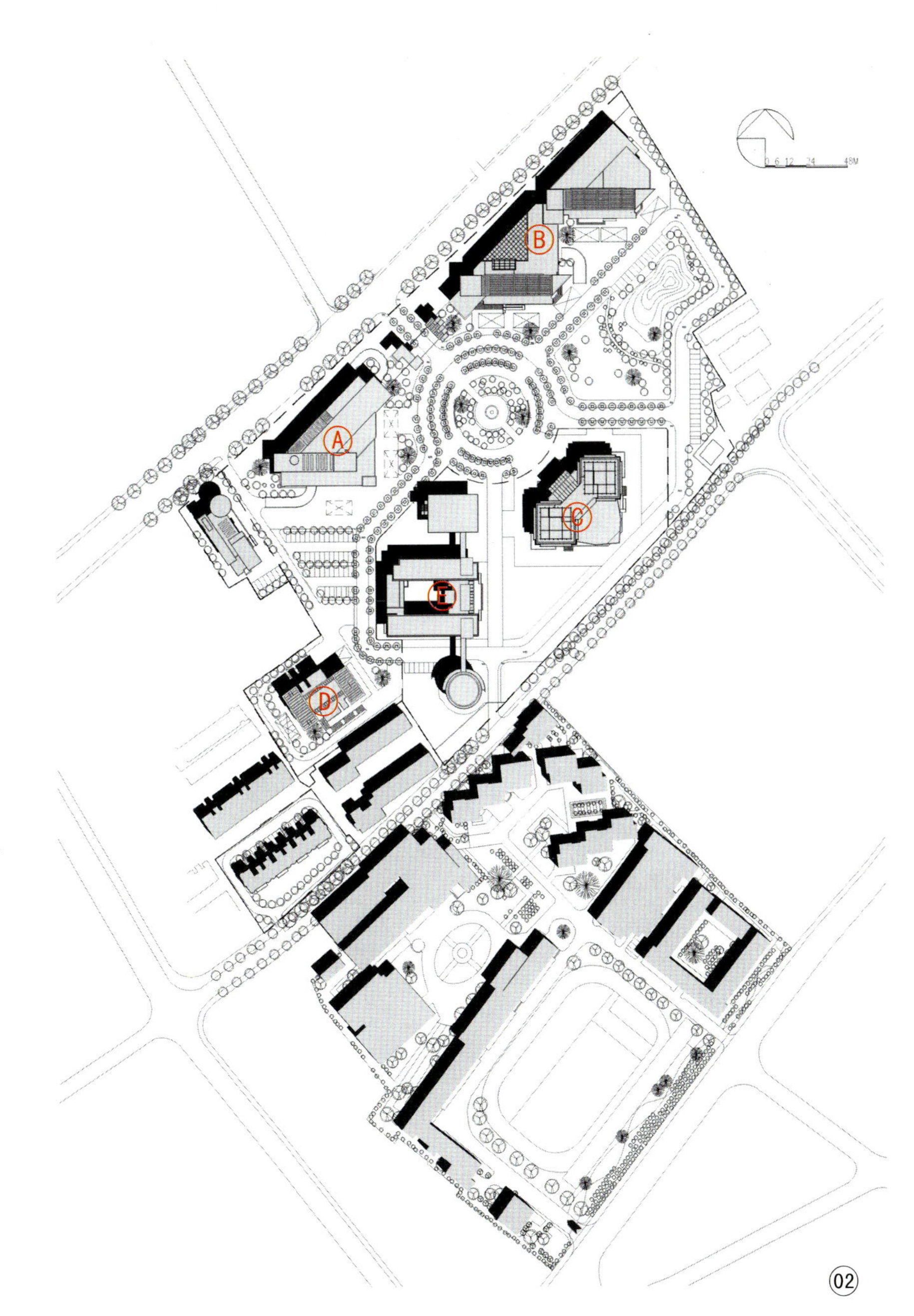

基本资料

地理位置：	上海市杨浦区
用地面积：	63,460m²
建筑面积：	84,520m²
其中	
图文信息中心：	16,488m²
教学综合楼：	9,760m²
国家电力南方培训中心：	21,105m²
电力科技楼：	10,693m²
电力研究中心：	13,273m²
研究生、留学生楼：	13,201m²
地上总建筑面积：	77,850m²
地下总建筑面积：	7,090m²
占地面积：	13,240m²
容积率：	1.22
绿化率：	38%
设计时间：	2005年

BASIC INFORMATION

Location:	Yangpu District, Shanghai
Base area:	63,460m²
Building area:	84,520m²
Among them	
Information Center:	16,488m²
Teaching Comple:	9,760m²
South Training Center of State Power:	21,105m²
Power Science Technology Tower:	10,693m²
Power Research Center:	13,273m²
Graduates and foreign students' dormitory:	13,201m²
On the ground total building area:	77,850m²
Underground total building area:	7,090m²
Site area:	13,240m²
Plot ratio:	1.22
Green ratio:	38%
Time:	2005

Analysis of Buildings

Information Center

New Information Center is a characteristic symbol of the new campus. As the tallest building, the 12-story tower stands straight against the main entrance and encircles the central round square. The tower leads the arrangement of the whole campus. Its simple, changeful shape makes it central building at the main entrance.

Teaching Complex

The teaching complex in the main axis, together with report hall in south and Library and

Information Center, encloses an entrance square in south part of the new campus. These changeful combinations of complex create a condition to build good campus atmosphere. The courtyard enclosed by the teaching complex not only provides a quiet place for teachers and students' communication after class but strenghen the spatial layers to build a transition space as well. These informal places for communication will attract a large number of students and scholars to chat, exchange and rest. The complex appears simple, modern and full of pace of times.

Power Science & Technology Tower

Located in northeast part of the campus, Power Science & Technology Tower consists of power system lab, power science and technology lab, electrical base lab and etc. In plane arrangement, Power Technology Tower interlaces with Power Research Center as a complex. We pay much attention to studying art and form in plane composition. The design reflects the reasonable combination of various functions. It emphasizes a close arrangement with abundant rhythms and changes. Therefore, it appears simple, free and changeful.

South Training Centre of State Power

South training center of State Power is located in northwest of the campus, along side Changyang Road. Its main entrance for vehicles is on that road. The main tower forms a folding shape in accordance with the spatial texture of the area and the planned axis of the new campus.

The design not only extends the space arrangement of the whole campus but also assorts with Changyang Road. Thus, it builds a good view along side the road. The main suites of the center are arranged in south-north direction. The podium stands parllel to Changyang Road, which appears simple, modern and harmonized with other buildings.

Power Research Centre

Power Research Center is located on west side of Electric power technology tower. There is an entrance inside the campus and the other at Changyang Road. The folding-line shaped building echoes the round square, which is the view sight toward the new campus seen from the old. Combined with some beams and slabs, the facade of the building has a strong visual impact and appears pure, simple and free.

Graduates and foreign students' dormitory

Graduates and foreign students' dormitory is a 15-story tower. Its close plane arrangement, simple and free shape with different heights reflects pace of the times in campus life.

03 形象完整统一的学院新区沿街面，呈现出电力学院新区新颖别致的建筑风格。

The facades facing the roads emerge as the same kind, which display its new and unique architectural style.

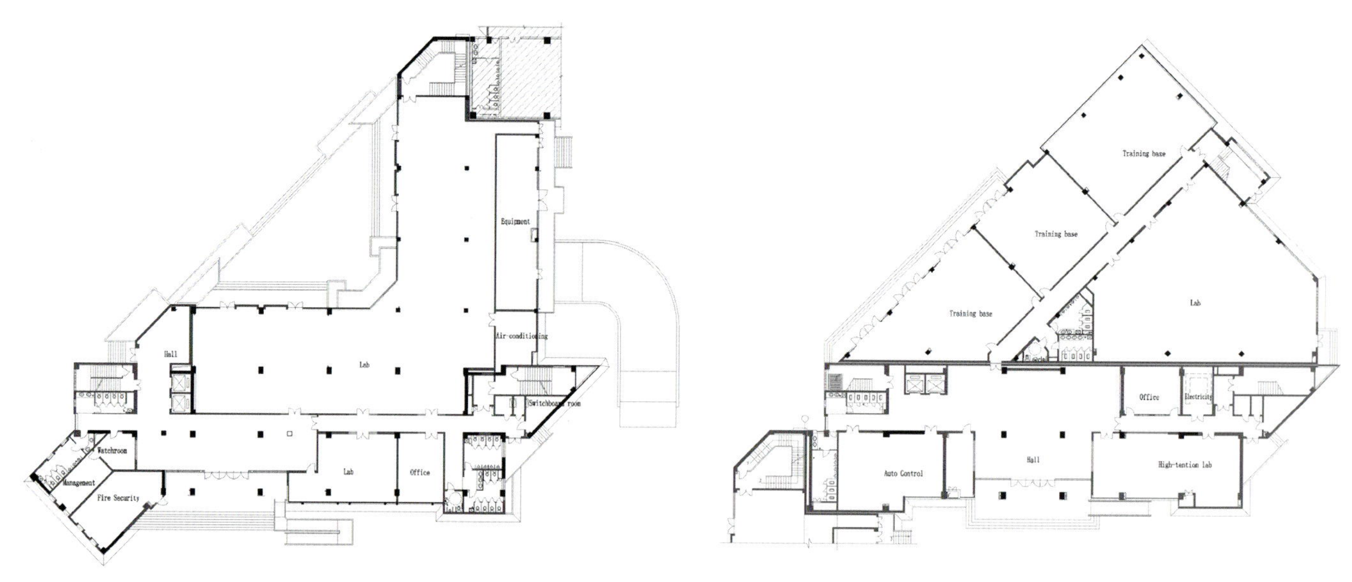

Fisrt floor / 研究中心一层平面

Fisrt floor / 科技楼一层平面

03

04 电力研究中心与电力科技楼面向校园核心广场区一侧，反映出典雅、安静的校园气氛。
Seen from the central square, Power Research Centre and Power Technology Tower reflect an elegant and quiet campus atmosphere.

04

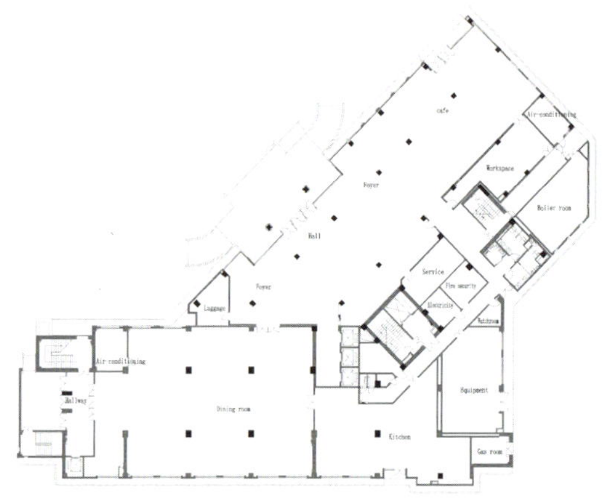

Fisrt floor / 培训中心一层平面

Facade/ 培训中心立面

Facade/ 培训中心立面

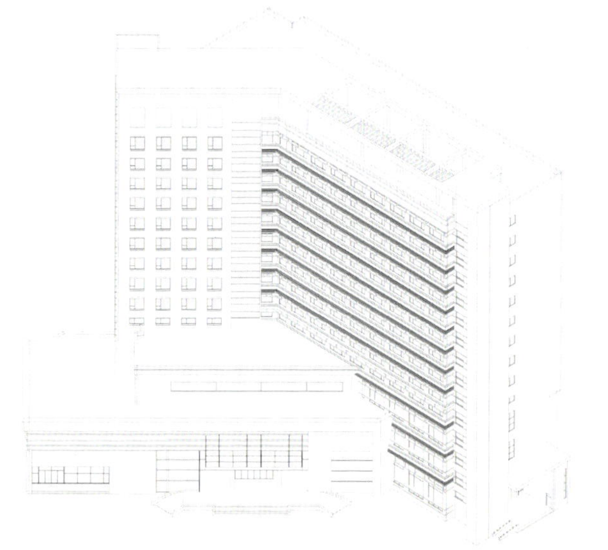

3D model/ 培训中心三维模型

05 国家电力南方培训中心实际上是一个四星级的酒店，但因为其所处的特定的校园背景，除入口大堂区以外，其主体建筑更多地与校内另两幢沿街建筑取得色彩，材质及形体关系上的协调。

Actually, South Training Centre of State Power is a four-star hotel. But due to its location, the Tower is more related to another two buildings in colors, material and shape along side the road , except for the lobby.

06 新校区夜景鸟瞰图，将六栋新建建筑所组合的新校区总体构架反映得清晰肯定。
Airscape at night, a clear picture of new campus which consists of six new buildings.

过程之美

——电力学院三思

从2004年底到2005年上半年，电力学院新校区的方案经历了几次反复的调整与深化，直至2005年5月初，我们开始正式启动这项工程。

新校区以圆形广场为核心，将新老校区的空间形态概念化为一根南北发展轴，针对小型校区采用了一条中外环线车道，将功能区分为科研、教学、生活、预留四个区片，这“一核、一轴、一环、四区”一起构成了紧密而有机的校区骨架系统，其中教学区主要由图文信息中心、教学楼组成，功能与老校区联系较为紧密，同属校园的内向型建筑；科研活动区主要由电力科技楼、电力研究中心组成，是上海电力学院产学科研合作和电力科技创新的基地，是对外交流的窗口，功能上属于校园的外向型建筑，可以通过他们展现学校的形象；生活服务区主要由电力南方培训楼、研究生留学生楼、专家楼、学生活动中心组成；而在中心广场东面预留一块完整用地以作为今后校园发展之用，目前作为校内绿化景观。

其中电力科技楼、电力研究中心、电力南方培训楼及研究生留学生楼作为一期工程首先开始进行方案深化及施工图设计。我有幸加入到这个项目组参与设计，写下一些随性的文字，记录设计的心情，体味设计历程。

创意

初识电力学院的建筑规划方案，对其仅有个大体的相对模糊的概念定位，并未深入地究其根源。随着深化设计的努力，编制完整的设计导则的出炉，我们逐渐接近了清晰的电力学院的新轮廓。电力学院的建筑应是什么样子的？尽管国际上有很多模式，具体到华谏，似乎有这样一些被固化了的意向：清雅，凝练，简约，极致。我们试图去迎合，但又不甘心被固化的思想所禁锢，思路也经历了一个阴雨蒙蒙到多云转晴的过程。在我们理想中的电力学院这组新建筑应该是：形体简洁有力，冷静挺拔，去掉一切不必要的突出之物，加入一切必要的细节，使其看似“简单”，实是“丰富”。

过程的开始

电力科技楼和电力研究中心是最先开始深入的两栋建筑，我们一边设计一边寻求着某种设计方法的契合。建筑创造的过程是在对立的两极间奋进而行，或者知之甚多，或者知之甚少；知之甚多倾向于自我模仿或模仿他人，这正是历史主义的陷井；而知之甚少则会不断寻求新颖与未知，产生出寻求未来是什么的动力，却又难于实现。正是这种已知和未知之间的对立，为我们进行建筑设计确立了一种特定的实践模式。并在已知与未知，经验与天真间找到融合，反复进行着平面与立面的协调设计。

刚开始我们设计过程中出现了手忙脚乱理不出头绪的状况，在平面与立面的调整中不断徘徊，始终不得其门而入，找不到让人满意的效果。经过反复摸索，我们似乎找到了一套较为得心应手的设计模式。首先在原方案的基础上进行形体的调整。掌握的原则是：只观其大体，不涉其细节。

在我们的设想中，一期建设完成后，沿长阳路一面的景观应全面打造完成，电力学院新校区在长阳路上的观瞻效果，应该体现其极富性格的一面，也同时是电力学院的新建筑给世人的第一面，必须有其新发现。

我们用电子模型模拟出电力学院沿街三栋建筑单体形体空间，着重研究电力科技楼和电力研究中心的形体，对预期的效果进行系统分析：如果形成平行于长阳路的铜墙铁壁将对城市景观形成压抑，并且使电力学院在城市中的外表黯然失色。因此一期建筑主体沿南北向布置，与长阳路形成45度的夹角，裙房沿长阳路展开，形成沿街丰富的建筑空间，高低对比，前后错落，空间通透，在长阳路上形成亮丽的风景。变换的视觉灭点让你对形体的尺度疏于判断，除了内心的景仰，也还有一丝依恋，并想追根究底地探寻个究竟，如此这般，便已达到了塑造这组建筑的目的。在这个大前提下，我们为了造就电力学院长阳路的“第一感观”，借用建筑形体的组合表达严密的逻辑思维特征，更加强调建筑自身的雕塑感，去其糟粕，留其精华，使形体的穿插更加简洁有力而无拖泥带水之物，使其突出退进之处的比例完美和谐恰到好处，更使其整体的形体达到一种效果——增之一分太肥，减之一分太瘦。

在形体调整完毕之后便可开始平面的调整，这其实是一个互动过程。任何建筑的首要条件都是解决其功能问题。而现代主义钢筋混凝土的受力体系本身造就了建筑不同于绘画、雕塑作品的随意性，在设计过程中我们尽量达到平面功能与立面形体恰到好处的融合一致，并精心考虑轴线、朝向、折线以及多重尺度等问题。

平面调整后便进行最后一步——立面的调整，这就关注到一个重要的方面——细节。

细节

设计和理解一幢建筑是不能与体现它的细部分开的，因为建筑细部，实际上就是研究整体和局部的关系，它相对于建筑和建筑空间来说是从属关系，它服务于建筑造型和建筑空间意境和气氛的表现。怎样进行细节设计？这是在设计中令人困惑的问题。其实细部概念并不是在详细设计阶段才出现的，我们在设计伊始就有可能触及细部。例如从电力学院的方案中可以感受到沿街三塔的韵律节奏、图文信息中心的秀丽挺拔、研究生楼的体量穿插以及教学楼的简洁有力，而为什么采取这种形象，又为什么选择这几种材料，这些都涉及了细部，当然具体的细部处理有可能会在此时，或是在后续的设计中被确定，但在构思时已经开始考虑它了。

而在立面调整阶段我们根据项目的投资、规模等会更加深入地考虑细部处理，通常我们在细节设计中把握以下原则：形态简约，宾主分明，繁简得体，材料适应，节奏鲜明，处理灵活。

首先是尺度与比例的划分及把握，我们分别在整体尺度、近人尺度和细节尺度上进行反复斟酌，尺度越大，把握越难，而比例的调整更是一种微妙的感觉，这无疑是一个不易解答的难题。

然后是考虑材料的肌理、光泽的合理搭配，精细而华美的立面细部均来自于对材料与肌理的编排与设计。这对于只在办公室画图的设计师来说是一种艰难的挑战。且不论我们对现在层出不穷的新材料根本一无所知，就连常用的玻璃、混凝土、涂料等材料的构造及特性都仅是一知半解。而材料商们为我们进行的产品推荐更使我们眼花缭乱，如雾里看花。

再者是对秩序与韵律的设计，就是在美学原则等的指导下组织门、窗、墙，点、线、面形成特定秩序的过程。这其实是一种感性认识，无法以某些特定地理论来描述它，当你在反复调整过程中突闪灵光，将某些设计元素放入了它原属之地，感觉是那么和谐舒服，如饮甘泉。有人也许会认为这是巧合，但我认为这是一个量变到质变的过程。

最后是色彩的确定，也是最难的一关。为什么我没有将它归在材料中，是由它的重要性决定的。在建筑设计中，色彩是最重要的造型手法之一，它是传达情感和创造气氛的重要元素，也是形成细部，强化细部的重要手段。确定色彩是我们最无法把握的，也是我最没有自信的。不纯净的色彩会使本应很出色的建筑变得平庸，不和谐的色彩会使我们的所有努力付诸东流。然而色彩的把握却是最不易的，它的丰富程度让你难以置信。不仅色彩本身多种多样，更因为色彩并不仅是其本身的颜色，而是人眼中的颜色，它是变化的。大块色彩与小块色彩的差异，一年四季与昼夜交替色彩的差异，镜面反射与漫反射色彩的差异，不同天气色彩的差异，甚至周边环境不同而造成的色彩差异等等，这些都使人在确定色彩时感到苦恼。

随着细节设计的完成，两栋自己也还满意的建筑跃然纸上。但设计还远远没有结束，对设计师来说，建筑没有竣工前，设计将一直伴随其而存在。

Lingang Campus, Shanghai Fisheries University

上海水产大学临港新校区

Shanghai, China, 2005.

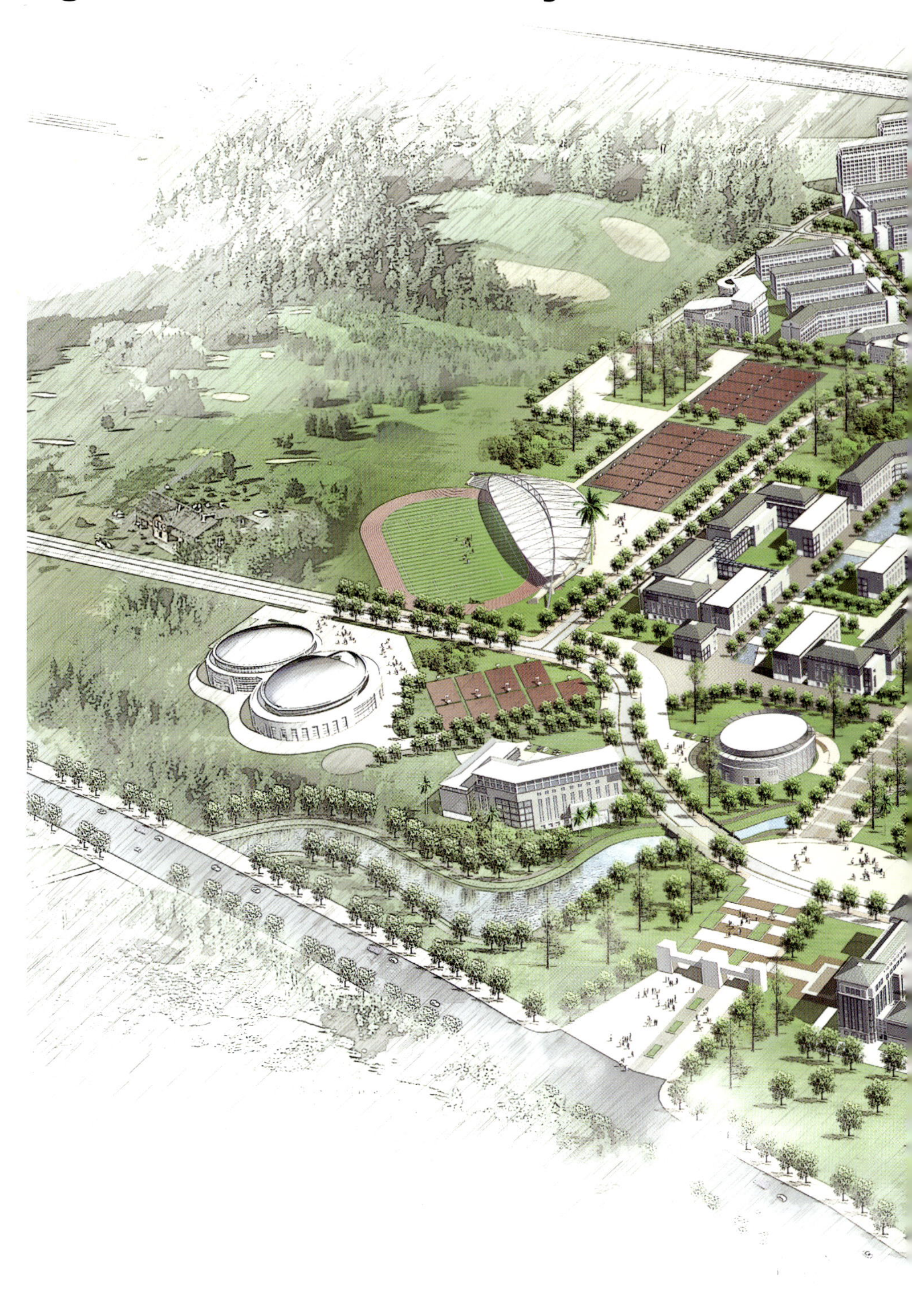

上海水产大学诞生于1912年，是中国历史最悠久的高等水产学府之一，被誉为“中国现代水产教育的摇篮”。今天的上海水产大学是一所以水产、海洋、食品等特色学科为主，农、理、工、经、文、管等学科协调发展的多科性大学，是国家水产科研、教育的重要基地。依据上海市高校布局结构调整计划，2008年水产大学需整体搬迁到远离市区的南汇洋山港后方基地——临港新城谋求发展。

水产新校区的建筑规划设计依托临港新城"生态城市、风景城市、旅游城市、数字城市" 的建设目标，注重水产大学历史人文气质与临港新城时代气息的融合。

在以人本主义为原则的指导思想下，新水产规划建筑设计以模块化设计理念为基础：即将校园建筑分成几大模块——教学楼模块、实验楼模块、学院楼模块、后勤模块、体育运动模块、公共行政模块。各大模块以合院为基础，形成群落。校区空间规划建立在分析各模块的内部关系结构及外界联系方式的基础之上。“中国传统建筑的精髓是幽深的庭院”在水产大学中有了完美的诠释：群落式的布局，递进合院式的空间。

“新水产”的风格化倾向是水产大学新校区又一重要特质，她放弃纯古典的或纯现代的表达方式，强调历史感与时代精神的结合；放弃形式的拼贴，强调意境的延伸；放弃大无边际、平铺直叙，强调平易近人、复杂而生动。这种倾向的背后带有一些江南水乡的气息，隐含一些传统书院的神韵，表现了一些海派建筑的洋气，又不乏高等学府文化底蕴。

上海水产大学临港新校区的启动有效地拉动了临港新城的城市文化产业综合竞争力，为新城建设注入了新的活力。

01 水产大学新校区总体规划鸟瞰效果图，注重表达了新校区规划设计的结构清晰性与逻辑完整性。

Airscape of SFU, reflects a clear structure and an integrative logic of the new campus.

Founded in 1912, Shanghai Fisheries University is one of the oldest fisheries institutes in China and is named as "the cradle of Chinese modern fisheries education". Now, Shanghai Fisheries University is a multi-fledged university committed to fisheries, ocean and Marine foods (its main feature) together with agriculture, science, industry, economy, art and management. SFU has become an important national base of fisheries science research and education. According to a new location adjustment plan of local universities in Shanghai, SFU will be moved to Lingang new town in Nanhui, the back base of Yangshan Harbor, for further development.

On the basis of the construction target of Lingang new town, 'an ecological town, a touring town, a digital town', the architectural planning and design of the new school buildings emphasize the integration and harmony between SFU's humanity and Lingang's development pace.

Directed by the principle with emphasis on humanity, the new campus is divided into several blocks including teaching, experiment, various schools, logistics, sports and public administration, based on the concept of blocking. Each block is combined with a courtyard and all yards make up a community. The spatial planning of the new campus comes from our analysis of interior relations and exterior connections of the blocks. The new campus of SFU with its communities and yards perfectly reflects the essence of Chinese traditional buildingswith the layer upon layer courtyard.

The style of 'New SFU' is another important character of the new campus. The new design abandoned the way of the pure classical or the pure modern expression instead emphasizing the integration of history and pace of times; abandoned the combination of modalities but focusing on the extension of atmosphere; abandoned a big, vacant and fade description but carrying out a close, complicated and active depiction. This style contains some verve of watertown and traditional academy, which indicates some modern flavor of buildings in Shanghai and the cultural atmosphere of academic institutes.

The construction of SFU Lingang Campus effectively strengthens the cultural industrial competence of Lingang new town, which pours into new vitality in urban constructions.

基本资料 / BASIC INFORMATION

基本资料		BASIC INFORMATION	
地理位置：	上海市南汇区临港新城	Location:	Nanhui District, Shanghai City
总用地面积：	1,068,000 m²	Gross Base area:	1,068,000 m²
总建筑面积：	586,000 m²	Gross Building area:	586,000 m²
其中：		Including:	
一期：	199,500 m²	1st phase:	199,500 m²
二期：	173,000 m²	2nd phase:	173,000 m²
三期：	47,000 m²	3rd phase:	47,000 m²
四期：	166,500m²	4th phase:	166,500m²
集中绿地面积：	302,800 m²	Concentrate green area:	302,800 m²
集中绿地率：	28.35%	Concentrate green ratio:	28.35%
水域面积：	61,634 m²	Wate area:	61,634 m²
水域覆盖率：	5.70%	Water overlay ratio:	5.70%
道路广场面积：	212,630 m²	Road & Square :	212,630 m²
容积率：	0.55	Plot ratio:	0.55
建筑占地面积：	153,187 m²	Site area:	153,187 m²
建筑密度：	14.34%	Building density:	14.34%
设计时间：	2005年	Time:	2005

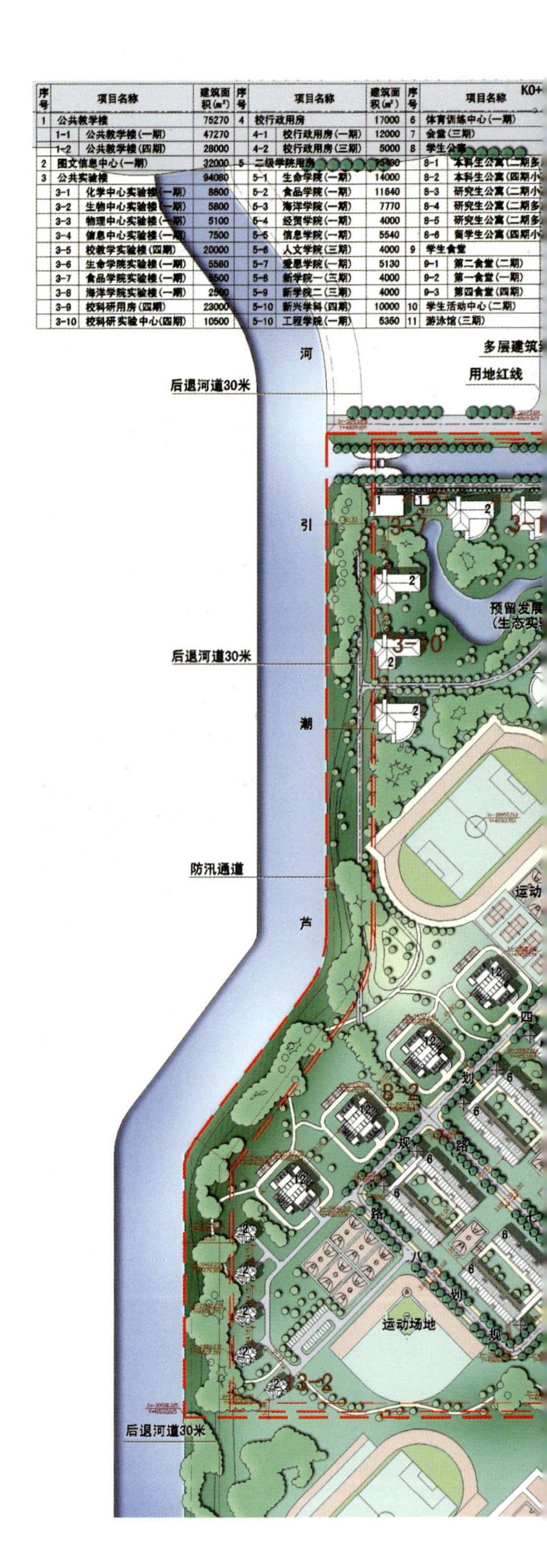

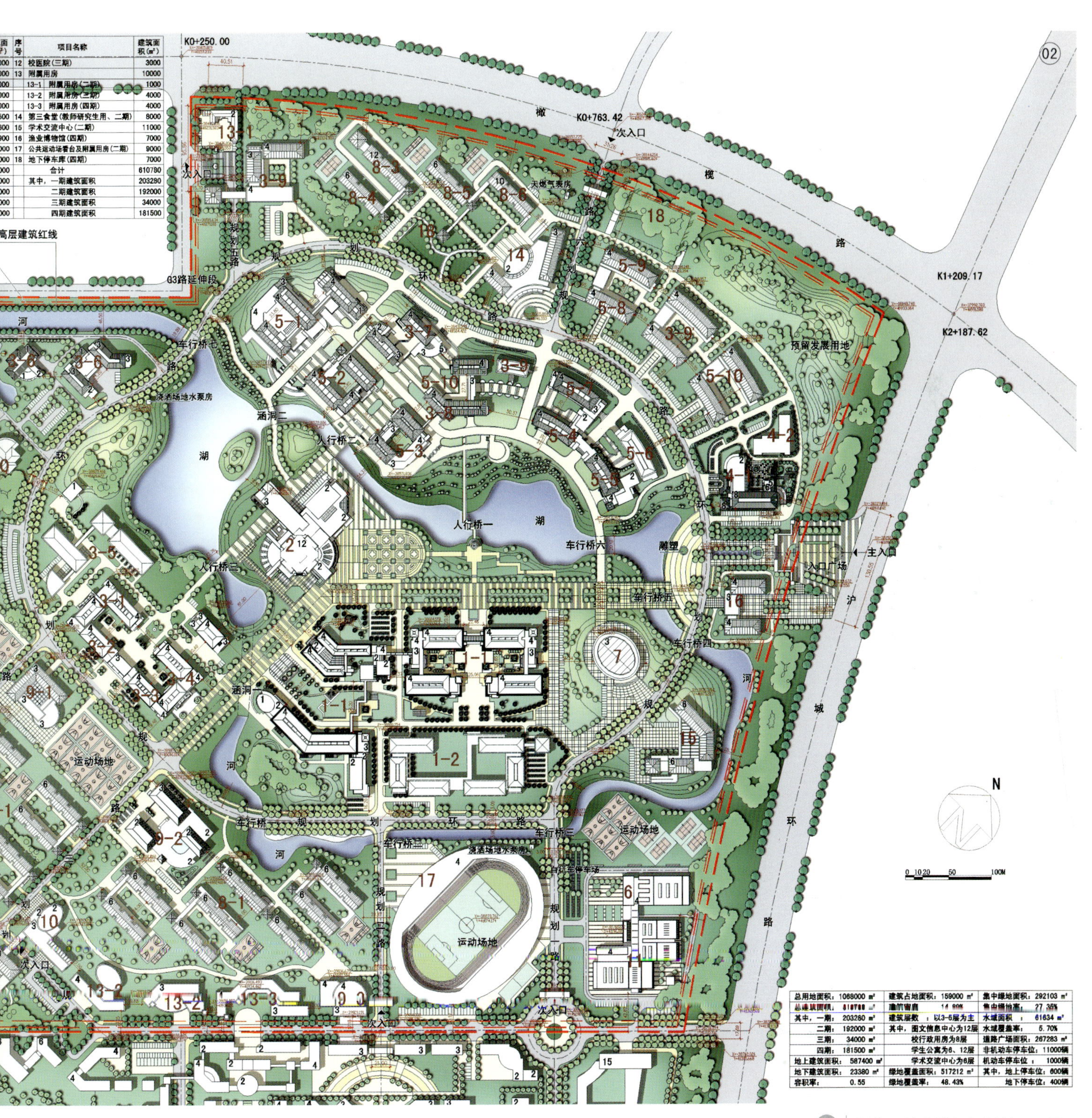

面积(m²)	序号	项目名称		建筑面积(m²)
000	12	校医院(三期)		3000
000	13	附属用房		10000
000		13-1	附属用房(二期)	1000
000		13-2	附属用房(三期)	4000
000		13-3	附属用房(四期)	4000
500	14	第三食堂(教师研究生用、二期)		8000
600	15	学术交流中心(二期)		11000
900	16	渔业博物馆(四期)		7000
000	17	公共运动场看台及附属用房(二期)		9000
000	18	地下停车库(四期)		7000
000		合计		610780
000		其中，一期建筑面积		203280
000		二期建筑面积		192000
000		三期建筑面积		34000
000		四期建筑面积		181500

总用地面积：1068000 m²	建筑占地面积：159000 m²	集中绿地面积：292103 m²
总建筑面积：[illegible] m²	建筑密度 14.90%	集中绿地率：27.35%
其中，一期：203280 m²	建筑层数 ：以3-5层为主	水域面积 ：61634 m²
二期：192000 m²	其中，图文信息中心为12层	水域覆盖率：5.70%
三期：34000 m²	校行政用房为8层	道路广场面积：267283 m²
四期：181500 m²	学生公寓为6、12层	非机动车停车位：11000辆
地上建筑面积：587400 m²	学术交流中心为6层	机动车停车位 ：1000辆
地下建筑面积：23380 m²	绿地覆盖面积：517212 m²	其中，地上停车位：600辆
容积率：0.55	绿地覆盖率：48.43%	地下停车位：400辆

02 新校区总体详细规划总平面图。

Detailed site plan.

A 图文信息中心

图文信息中心位于校园的核心部位，三条步行主景观带的交汇处。北为院系楼组群，南倚实验楼，西临校园景观湖，东南侧为教学楼群。

图文信息中心建筑形式的确定源自对规划精神的理解和对环境的分析，其平面布局与空间形态体现了轴线的起承转合关系和对校区全局的统领作用。

主楼一层底部设置自修教室和自修阅览室，二至十一层为图书馆，其中二层为现刊阅览室，三到六层为基本书库，七、八层为过刊阅览室，九至十一层为高科技阅览室及领导办公室。校史馆与档案馆设于西北侧临湖的裙房中，南侧裙房中设有现代信息与教育中心，北侧裙房为会议中心。

主楼造型俊秀挺拔，对传统三段式的现代表达方式以及中国传统与西方古典建筑语汇的结合方式进行了探索。

B 行政会议中心

行政会议中心位于预留发展用地和规划环路之间，西接二级学院用房，东临校园主入口的入口广场，是学校主轴线的起点。

行政会议中心主楼八层，裙楼二层，以校园的日常办公为主，兼有会议和接待功能。主楼内部空间以内廊组织，空间紧凑；三至六层局部挑空，增添了变化。校领导办公室设置在建筑的七八层，朝向校园入口广场，可俯视整个校园，景观良好。裙楼主入口设计了两层通高的门厅，作为人流集散和形象展示的空间。与门厅相连的是一个休息空间，朝向半开敞式内庭，作为等待区，同时提供一个舒适的交流场所。

C 教学楼群

教学楼群位于学校规划环路和主轴步行道围合的地块中，共四栋教学楼，采用庭院式空间布局组织联系。教学楼内部建立有完整的步行系统，以连廊作为支撑步行活动的骨架。在景观设计上，将台地、片墙、树阵、水池、草坡、情景雕塑等与交通有机组织，功能性与景观性相得益彰，凸现出场所精神，形成校园最人性化的空间。

立面设计上强调中国传统与西方古典元素的结合，采用“传统三段式的现代表达”，既内敛稳重、又不失挺拔刚毅。色彩以多层次的灰色为基调，其间点缀黑色钢结构构件，并适时以彩色片墙和深色花窗作为点睛之笔，营造出清幽雅致的校园氛围。

D 教学实验楼

教学实验楼位于新校区主景观湖南侧，基地北面为校图书馆，南临规划环路，远望学生宿舍楼，东面为教学楼组群。

实验楼组群由 4 个专业实验中心、学生事务中心及危险品仓库组成。实验中心平面布置以 8 m 柱网为模数，设置各类实验室，空间灵活多样，适用性强。

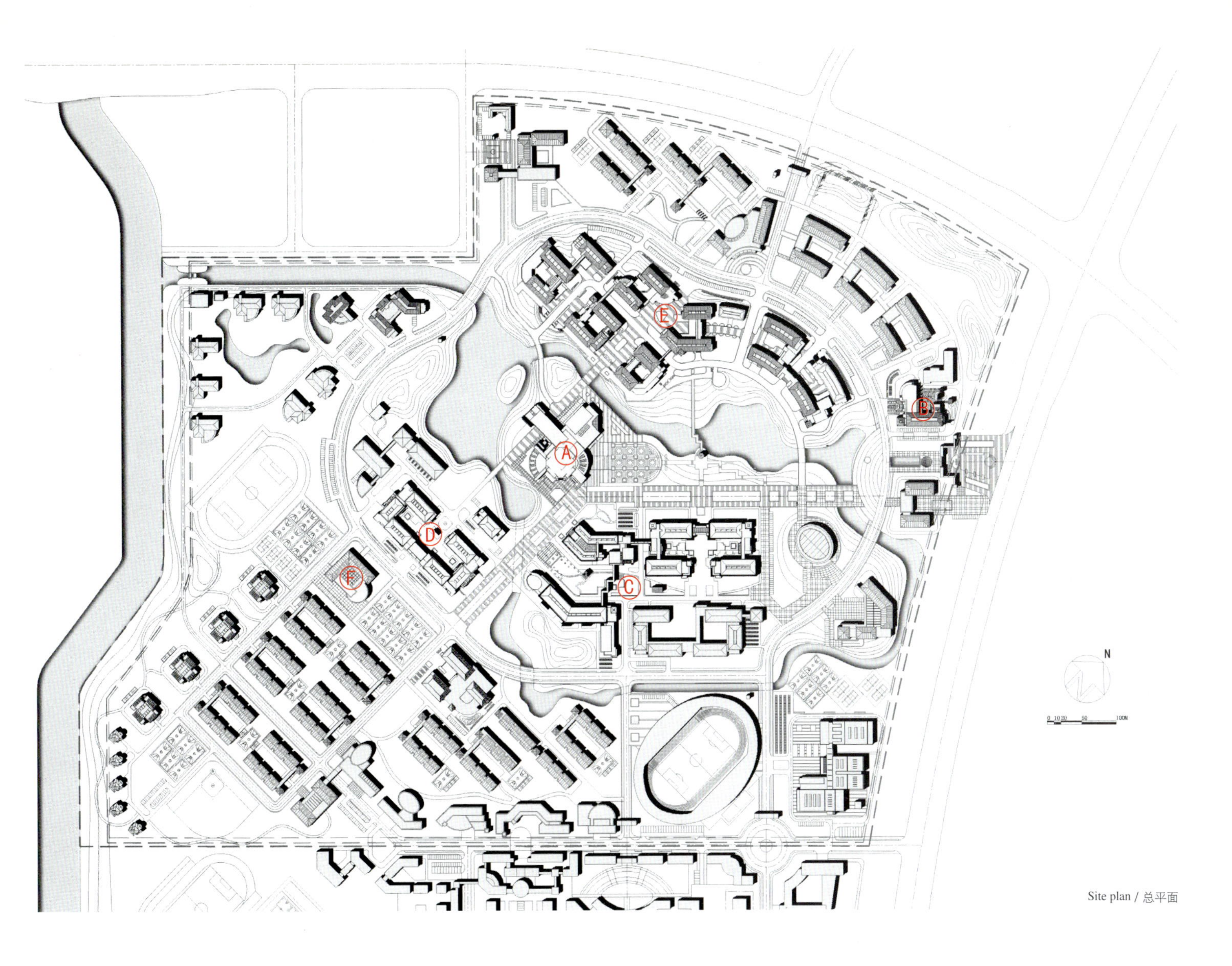

Site plan / 总平面

E 二级学院楼群

二级学院楼群位于校园中轴大道右侧，东倚校园中心湖面，与教学楼隔岸相望；南临图文信息中心，西北、东北侧分别为研究生公寓与行政用房。

二级学院楼群由生命学院、食品学院、海洋学院、工程学院、经贸学院、信息学院、爱恩学院、校科研用房八部分组成，为各学院的教学科研办公用房 。

国人常用“人杰地灵”来概括自然环境对人的陶冶作用，二级学院楼群的塑造，即体现了华谏国际对环境营造的注重。设计以自然为范本，追求“虽由人作，宛自天开”的境界，“山水意境、书院情怀”，是其建筑性格的写照。其多变的竖向场地和丰富的庭院空间，充分的展示了场地的自然之美，体现了尊重自然，强调与自然和谐共生的设计理念。华谏国际在水产大学创作中追求的境界——朴素的建筑观、寄情山水的环境观、追求意境的人文观、书院建筑空间组织的礼乐精神——在二级学院设计中得到了完美展现。

F 学生食堂

学生食堂北邻校园环路，与公共教学区遥相呼应，南面、东面为本科生生活区，西面为活动区。基地位置处于教学区与生活区之间的枢纽地带，保证了各个方向人流到达的便捷性，提高了学生的使用效率，满足总体规划要求。

食堂整体为二层，局部为三层，可供1052人同时就餐。学生食堂内还设有2,000 m^2学生生活广场，安排有商业和生活服务。生活广场与食堂分共同围合成一个内凹的院落空间，创造了真正意义上安静、舒适的生活广场，同时也是对规划整体脉络的延续和继承。

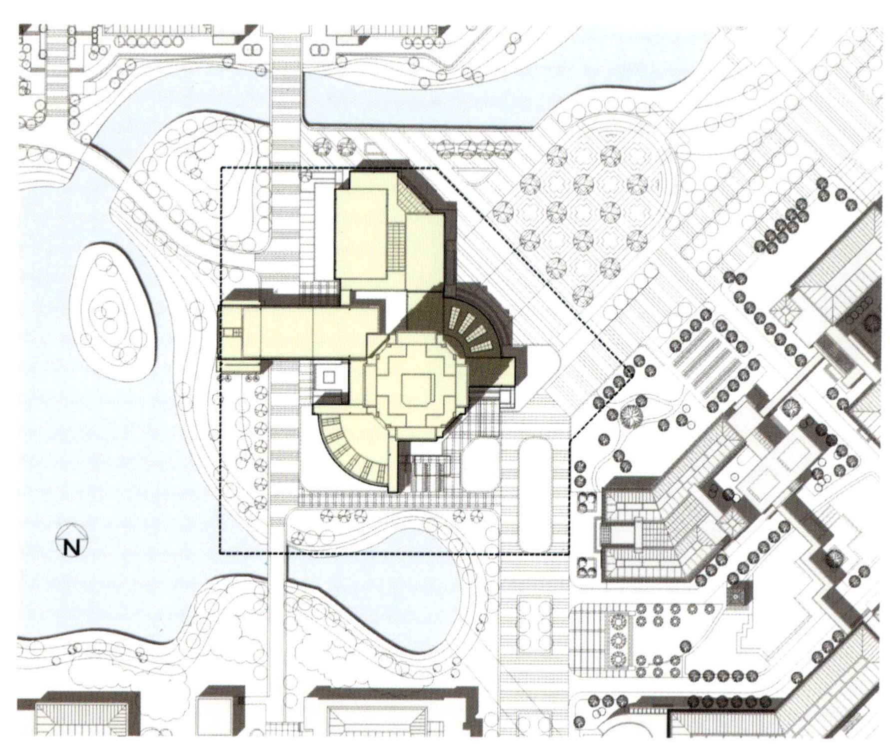

Site plan / 图文信息中心总平面

基本资料

用地面积：	22,024m²
总建筑面积：	32,189m²
建筑占地面积：	5,408m²
容积率：	1.46
建筑密度：	24.6%
绿化率：	40%
设计时间：	2005年

BASIC INFORMATION

Base area:	22,024m²
Building area:	32,189m²
Site area:	5,408m²
Plot ratio:	1.46
Building density:	24.6%
Green ratio:	40%
Design time:	2005

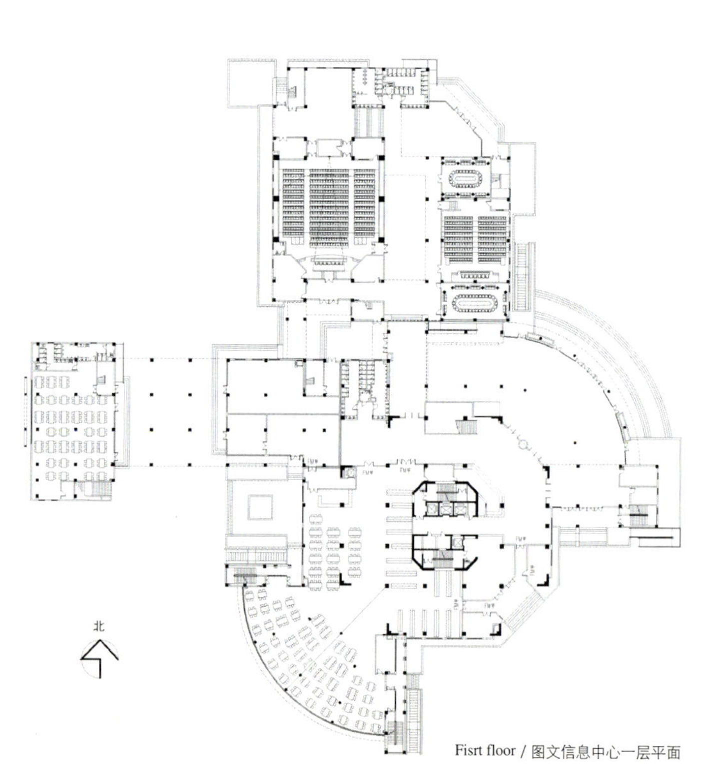

Fisrt floor / 图文信息中心一层平面

Library & Information Centre

Library & Information Center is located in the intersection of three main green belts and the center of the new campus. It stands at the south of various schools and colleges buildings, north of experiment building, east of the main lake and northwest of teaching complex.

The architectural style of Library & Information Center originates from the understanding of planning and the analysis of environment. Its plane arrangement and spatial modality reflects the important roles of axis: the leadership of the whole campus and the connection between the preceding and the following.

There are self-study rooms and reading rooms on the 1st floor of the main tower. The 2nd to 11th floor serves as the library, where the 2nd floor is the reading room for present periodicals and newspapers; 3rd to 6th floor are stack rooms; 7th to 8th floor are the reading rooms for past periodicals and newspapers; 9th to 11th floor are curator' office and the reading rooms for high-tech books. In podium, southeast of the lake, lies the scholastic history hall and document hall. While modern information and education center is in south podium and the convention center is in the north.

The main tower looks grand and straight. Its design is an exploration of traditional style of trisegment and a combination between Chinese traditional architecture and western classical architecture.

03 高层的图文信息中心处在校园主轴线的转析点上，是校区的核心建筑，总体体量竖向挺拔，横向舒展，形象典雅大气，且有浓郁的书院气息。

As the key building on campus, the high-rise is standing at the turning point of the main axis. It extends to a big scale both in horizontal and vertical space. It looks elegant outside and has an atmosphere of academy.

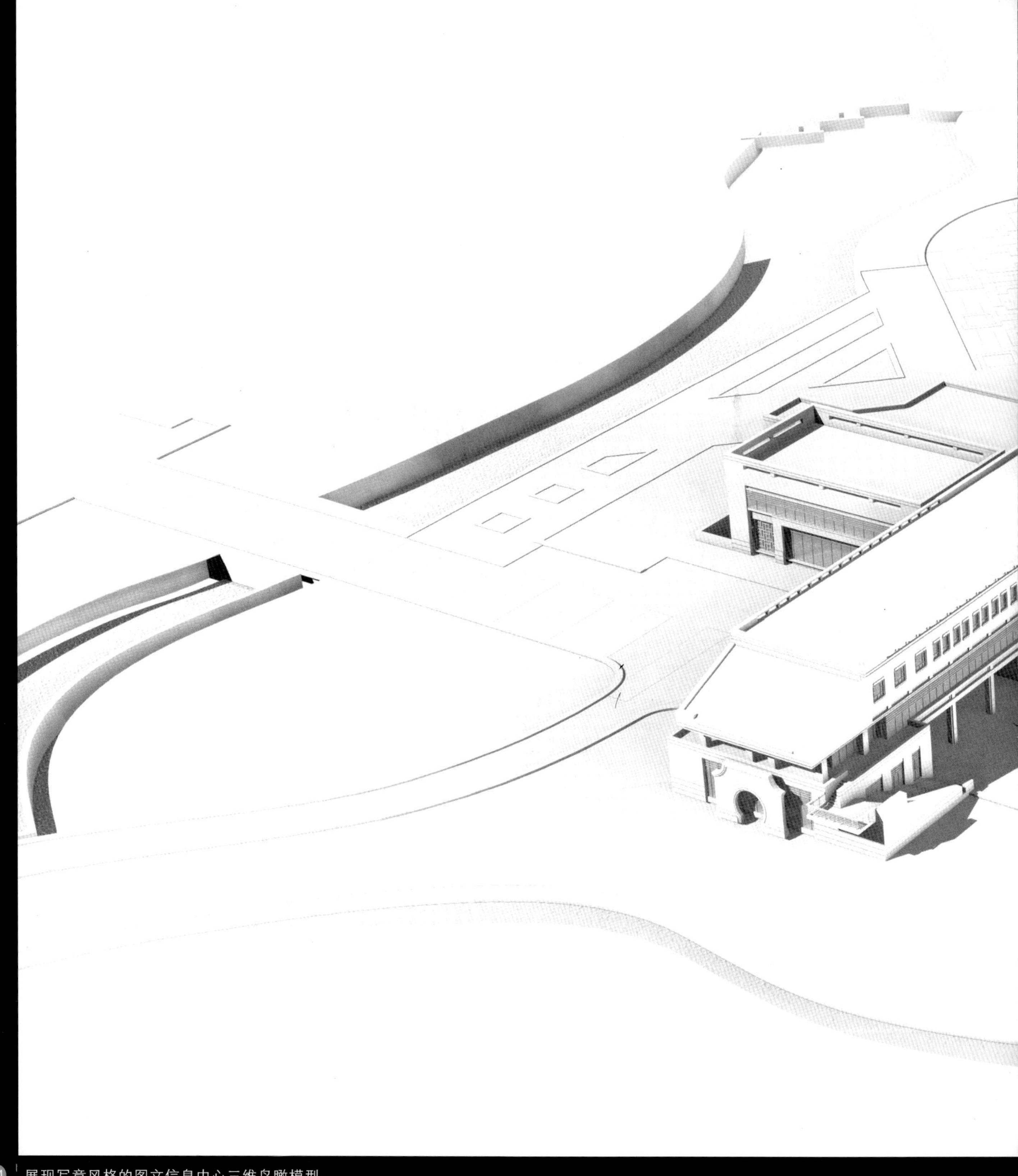

04 展现写意风格的图文信息中心三维鸟瞰模型。
Airscape of 3D model, reflects an enjoyable and comfortable atmosphere of Information Centre.

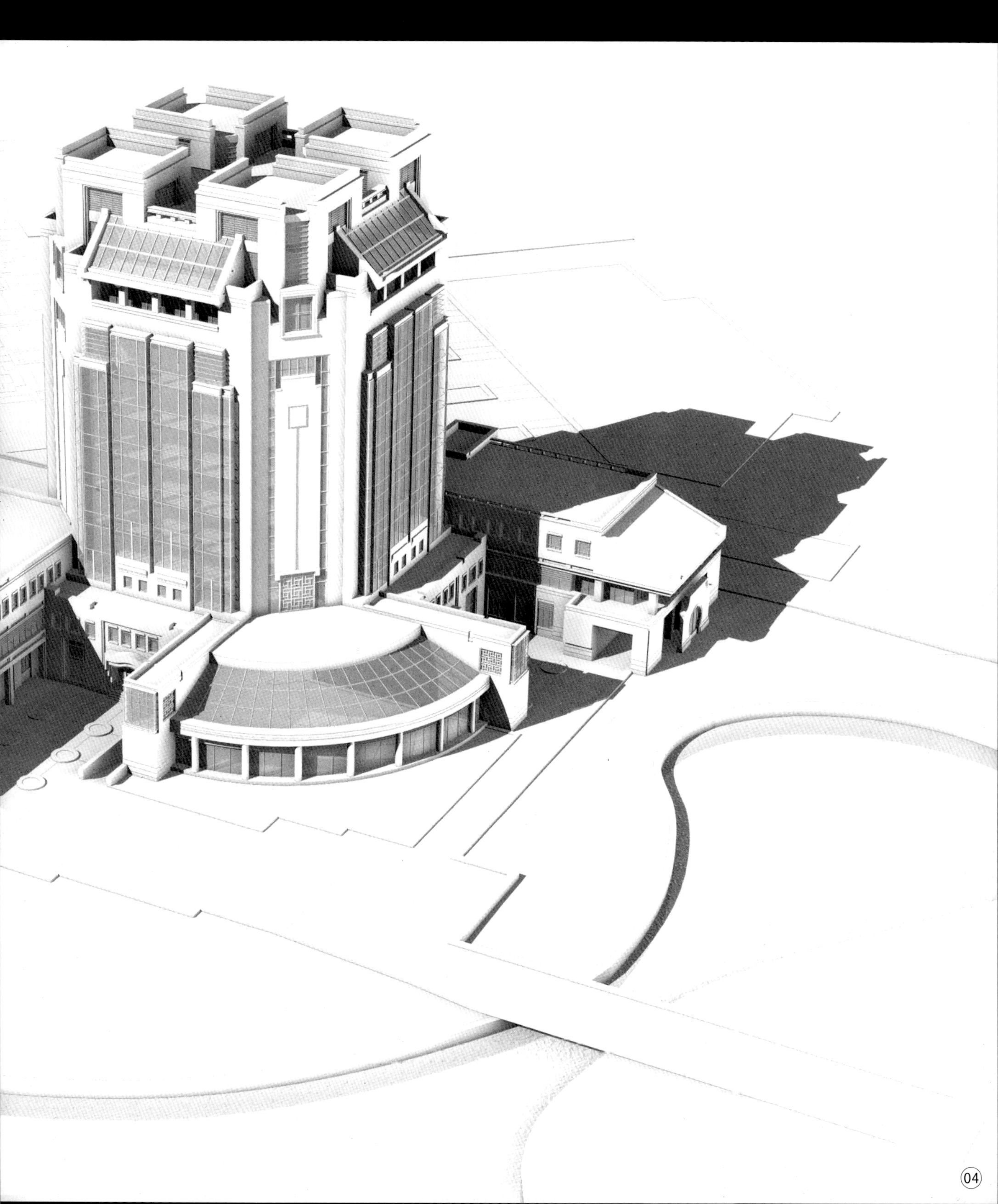

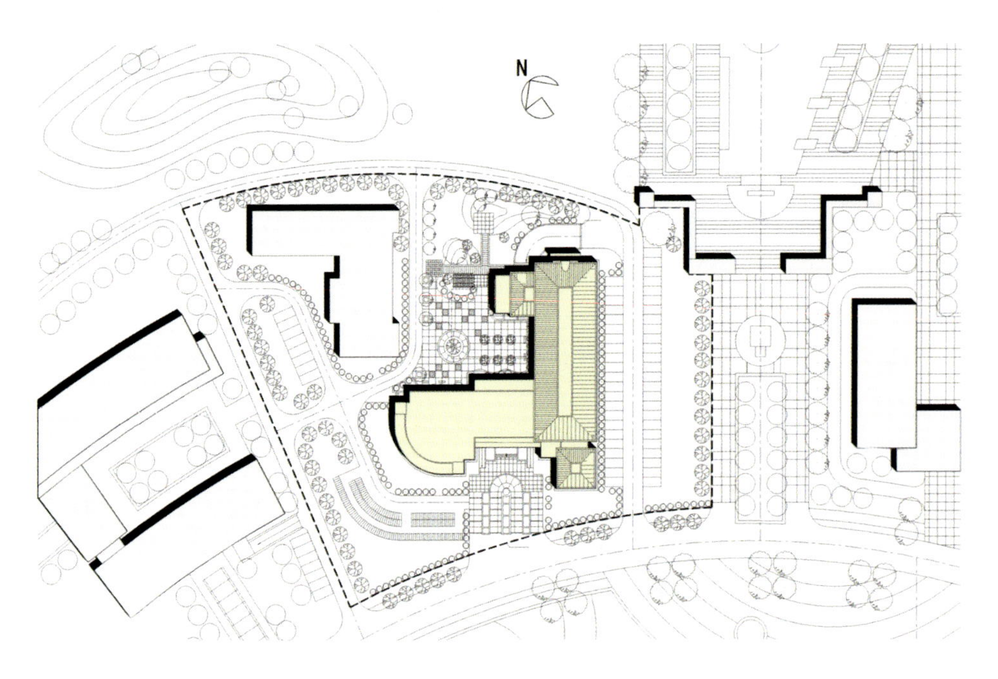

Site plan / 行政会议中心总平面

基本资料

用地面积：	14,415m²
总建筑面积：	11,610m²
建筑占地面积：	1,658m²
容积率：	0.81
建筑密度：	11.5%
绿化率：	37.28%
设计时间：	2005年

BASIC INFORMATION

Base area:	14,415m²
Building area:	11,610m²
Site area:	1,658m²
Plot ratio:	0.81
Building density:	11.5%
Green ratio:	37.28%
Design time:	2005

Administration & Convention Centre

Administration and convention center is located between the reserved land and the planned circular road, east of college complex, west of the main entrance. The center is the beginning point of the main axis of the campus.

Administration and convention center consists of an eight-story tower and a two-story podium. It serves for daily administration, management, convention and reception. Corridors with compact space organize the inner space of the main tower. It becomes changeful that part of the slabs from 3rd to 6th floor is removed. The leaders' offices are set on 7th and 8th floor toward the entrance square, which have a good view of the whole campus. The two-story hall at the entrance of the podium is designed as a space for stream organization and image display. A rest place is connected with the entrance hall. As an area for waiting, it is half open to the courtyard and provides a comfortable place for communication.

05 8 层高的行政会议中心是比较难处理的建筑，而其精巧的比例推敲显示了设计师深厚的功力，其中一个 8 层高的塔楼与图文信息中心塔楼形成一大一小的呼应。

The eight-story Administration & Convention Centre is not easy to grasp. Its accurate proportion denotes the capability of the designer. The small eight-story tower inside echoed the big information center in the distance.

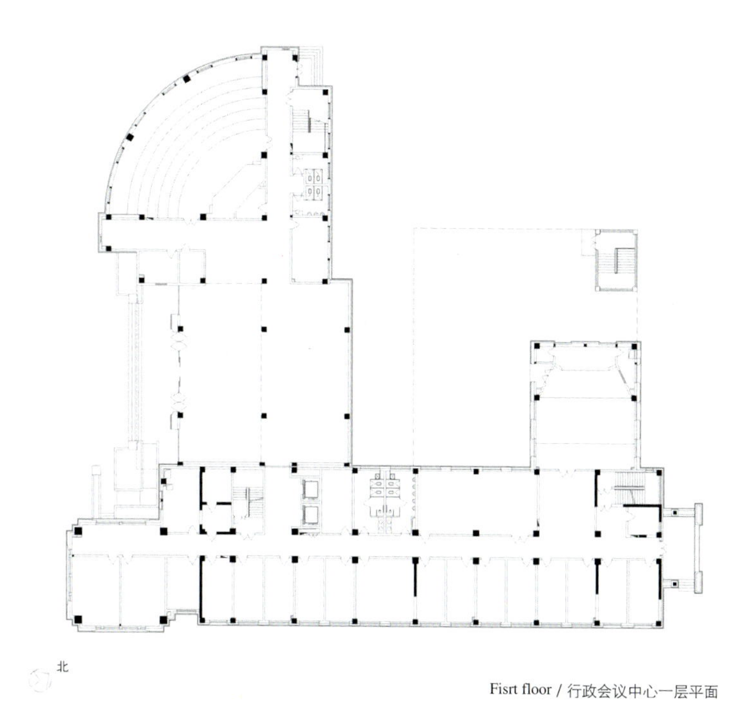

Fisrt floor / 行政会议中心一层平面

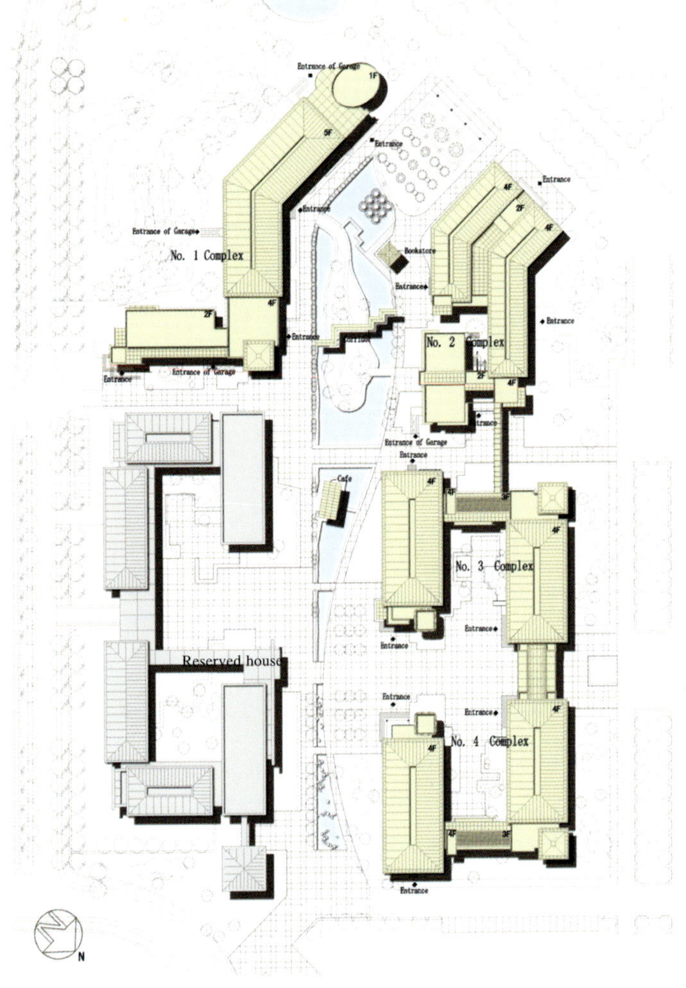

Site plan / 教学楼群总平面

基本资料	
用地面积：	68,486m^2
总建筑面积：	47,139m^2
建筑占地面积：	12,621m^2
容积率：	0.65
建筑密度：	8.4%
绿化率：	35%
设计时间：	2005年

BASIC INFORMATION	
Base area:	68,486m^2
Building area:	47,139m^2
Site area:	12,621m^2
Plot ratio:	0.65
Building density:	8.4%
Green ratio:	35%
Design time:	2005

Teaching Complex

The teaching complex lies in the block enclosed by the main walking streets and planned circular road. It consists of four teaching buildings connected by courtyards in spatial arrangement. There is an integrated walking system, which is made up of corridors inside the complex. In landscaping design, grounds, pieces of walls, arrays of trees, water ponds, slopes of grass and sculptures are organically integrated with traffics. Its function and scenery is well combined and set off each other. Thus, it forms the best place with humanity on campus.

In facade design, we emphasize the combination with Chinese traditional and Western classical elements. We use traditional style of trisegment to represent an introversive, steady, forceful and gritty state. Based on colors of different grays, the facade is decorated by black steel structures and properly interspersed with colorful pieces of walls and fuscous grilled windows as highlight so as to create elegant and quiet campus atmosphere.

06 教学楼多为三四层的建筑，其多栋单体所形成的书院空间的布局是最大特色。

Most of the teaching buildings are four or five stories; the courtyard of academic atmosphere that they encompass is the most important feature.

教学楼群内部所围合成的院落或大或小、情趣丰富，有开敞、有围合，显出了传统书院丰富的庭院空间表情。

The courtyards inside the teaching complex, which are big or small, open or closes, display various spatial expressions of traditional academy.

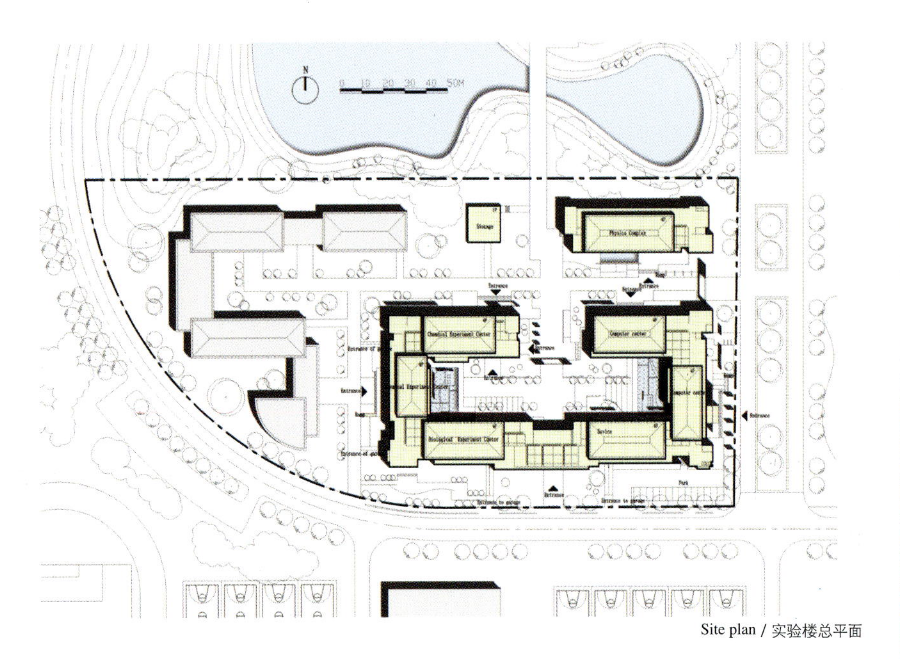

Site plan / 实验楼总平面

基本资料		BASIC INFORMATION	
用地面积：	39,380m²	Base area:	39,380m²
总建筑面积：	27,131m²	Building area:	27,131m²
建筑占地面积：	7,569m²	Site area:	7,569m²
建筑密度：	19%	Building density:	19%
容积率：	0.69	Plot ratio:	0.69
绿化率：	62%	Green ratio:	62%
设计时间：	2005年	Design time:	2005

Experiment Building

Experiment building is located to the south of the main lake and the library of the new campus, north of the planned circular road, next to students' dormitories and west of teaching complex.

Experiment building consists of four professional experiment center, a students' affairs center and storages for the hazards. There are various labs in the building. The columns divide the experiment center by 8 meters in plane arrangement, which makes the space more flexible and adaptable.

09 实验楼群紧临校园南北向的次轴，因其合院式的布局自然形成一个穿过图文信息中心裙房的次轴空间序列。

The experiment building is located near the main axis of south-north, naturally creating a sub-axis space passing through the podium of Information Centre, due to its enclosing arrangement.

10 实验楼的南向主立面注重对传统建筑形态、几何化、抽象化的表达。

The south facade of the experiment building emphasized an geometrical and abstract expression on traditional architectural modality.

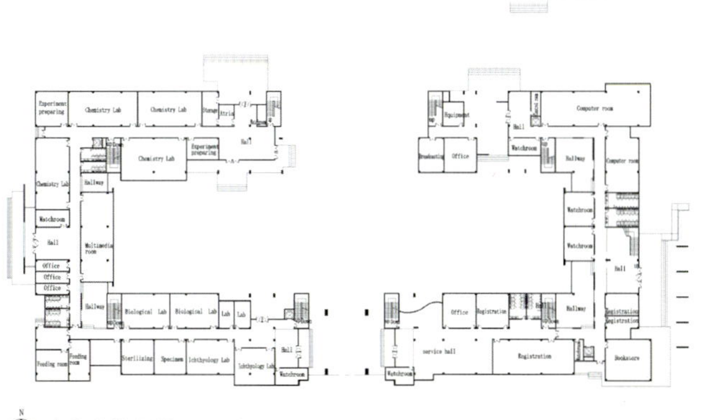

Fisrt floor / 实验楼一层平面

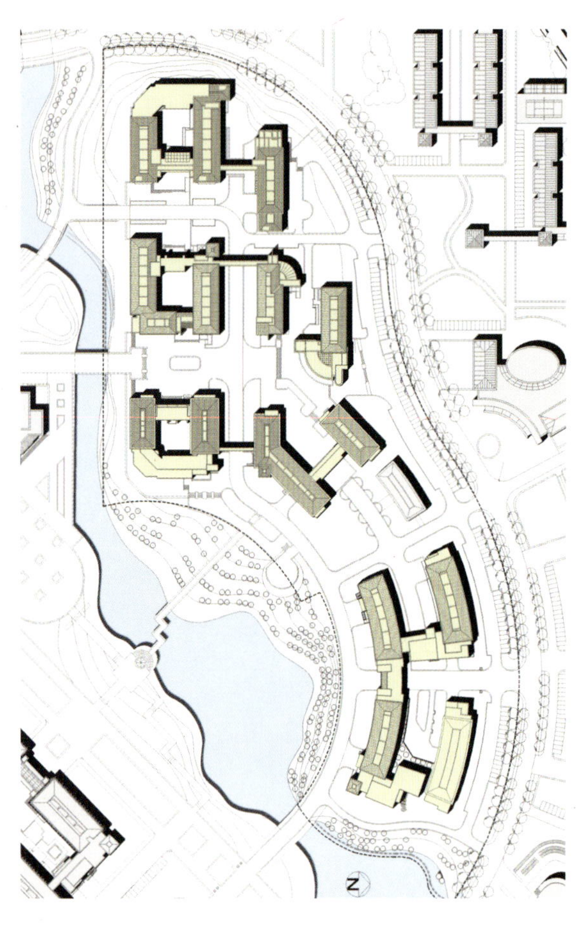

Site plan / 院系楼群总平面

Complex of Colleges

Complex of colleges is located on the right side of the main avenue, west of the main lake and next to teaching buildings, north of library and information center, southeast of graduates' dormitories and southwest of administration houses.

Complex of colleges servers as management houses for education and scientific research of different colleges including College of Aqua-life, College of Food Science, Marine College, College of Economics and Trade, College of Information, IEN Institute and etc.

Chinese people always cite 'outstanding person comes from great place. ', which indicates that natural environment plays an important role of people's life. The construction of complex of colleges reflects that DHGP pays much attention to building good environment. The design pursues the idea of following nature and attempts to create a pure natural environment inside the building. Its changeful vertical grounds and interesting courtyard space shows the natural beauty of the ground and the design concept of respecting the nature and living harmoniously with the nature.

基本资料

基本资料		BASIC INFORMATION	
用地面积：	82,856m²	Base area:	82,856m²
总建筑面积：	69,891m²	Building area:	69,891m²
建筑占地面积：	23,862m²	Site area:	23,862m²
生命学院建筑面积：	14,842m²	Collage of Aqua-life:	14,842m²
食品学院建筑面积：	12,834.5m²	Collage of Food Science :	12,834.5m²
海洋学院建筑面积：	12,679m²	Marine Collage:	12,679m²
信息学院建筑面积：	5,797m²	Collage of Information:	5,797m²
经贸学院建筑面积：	5,047m²	Collage of Economics and Trade:	5,047m²
爱恩学院建筑面积：	5,374m²	IEN Institute:	5,374m²
校科研用房建筑面积：	13,318m²	Houses for SFU scientific research:	13,318m²
建筑密度：	28.8%	Building density:	28.8%
容积率：	0.87	Plot ratio:	0.87
绿化率：	36.8%	Green ratio:	36.8%
设计时间：	2005年	Time:	2005

11

11 二级学院地处校园核心绿化景观区右侧地块，建筑走向成带型布局，各院系楼依次排开，完成对核心绿地的环形环抱，同时在楼群内部组织成一条由院落空间所围合成的庭院走廊，将各院系楼串联成一个完整的整体。
The complex of collages is located on right side of the central green area on campus. Different buildings connect each other as a belt and encompass the central green area in plane arrangement.

1.2 学生食堂在设计上更强调对细节的精准推敲。
We emphasize accurate design in details in the canteen.

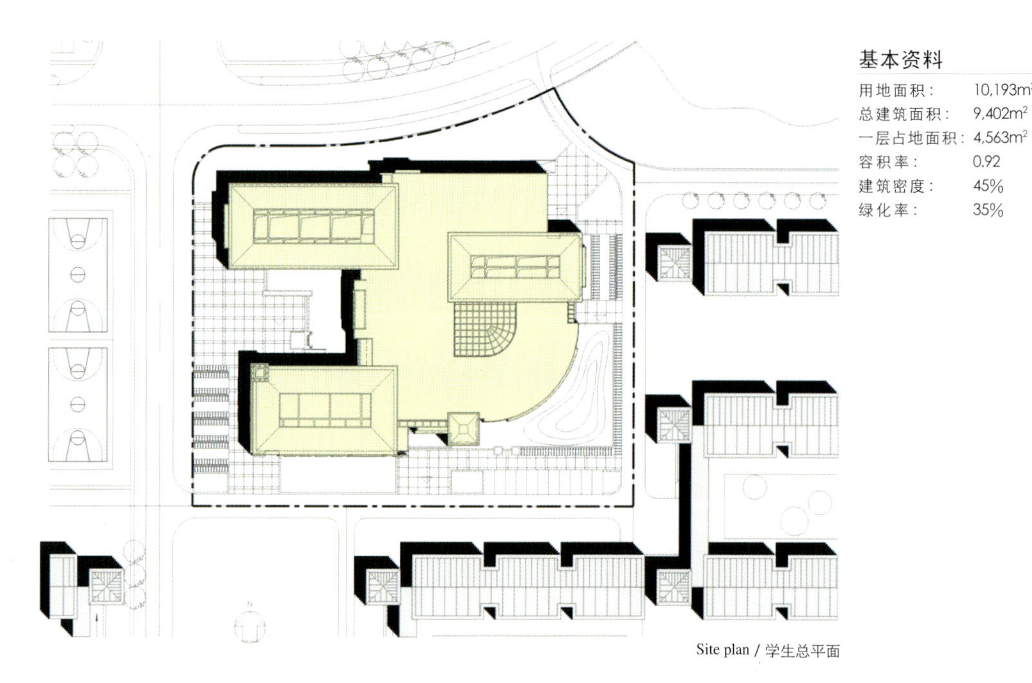

Site plan / 学生总平面

基本资料	
用地面积：	10,193m²
总建筑面积：	9,402m²
一层占地面积：	4,563m²
容积率：	0.92
建筑密度：	45%
绿化率：	35%

BASIC INFORMATION	
Base area:	10,193m²
Building area:	9,402m²
First site area:	4,563m²
Plot ratio:	0.92
Building density:	45%
Green ratio:	35%

Students' Canteen

Surrounded by living area and activity area, the students' canteen is located at south of the circular road, next to public teaching area. The canteen is standing between the living area and the teaching area so that it provides convenience and efficiency for students to all directions. Thus, it meets the requirements of the master plan.

The canteen is a two-story (three-story in part) building. It can provide food to 1,052 people at the same time. There is also a students' living square of 2,000 square meters with commercial and living service in the canteen. The living square and the canteen form a concave courtyard space and create a real quiet and comfortable place, which is also an inheritance and development of the master plan.

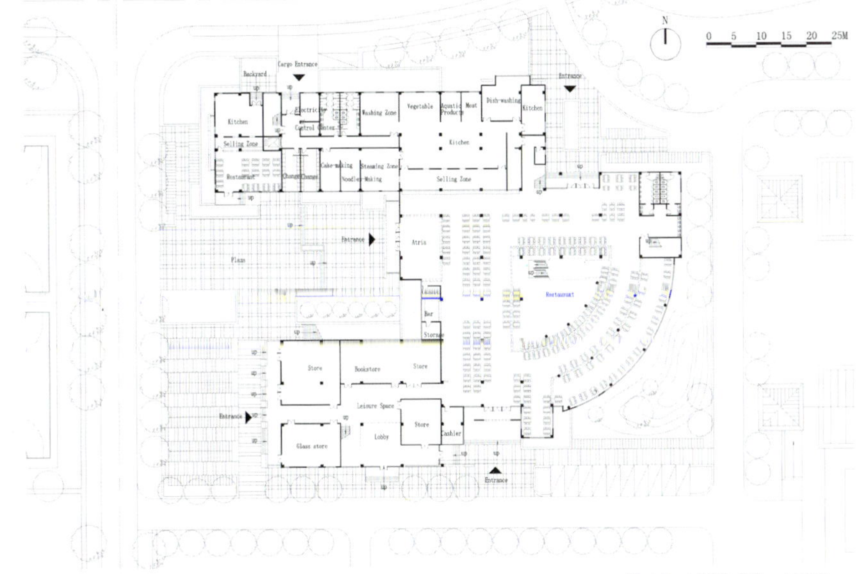

Fisrt floor / 学生食堂 一层平面

让大学精神在这里活下去

——诗话水产

“多才雅得江山助”，怡情悦性的校园环境是学生兴风花雪月的雅兴、立修齐治平之大志的寄托与激励，是他们文思泉涌的灵感源泉。纵观古今中外的大学发展历史，无不注重校园环境教化育人的重要作用，学校的文化传统通过校园空间、建筑风格以及环境小品的细节设计得以阐释和展现。而校园环境空间因其独特的人文内涵往往能成为一个大学校园的象征。

胡兰成在“随笔六则”中谈到：“以前到过的名胜印象都很淡，倒是常走的小街小巷对我有感情。我游过西湖，见过长城，可是动人的只是当时的情景，不是当地的风景。”同理，当人们渐渐远离校园的宁静，融入城市的喧嚣之后，再回忆起当时的校园，也许也会有一丝感动，让记忆停留在某个情景的片断，充分感受校园的纯净。“记忆”既不同于史实，也不同于虚构，它应该怎样度量，又怎样外化呢？怎样给我们的校区赋予这一种亲切的情感？

一个大学的建立和发展有着它特有的历史背景和文化渊源。历史感使一切建筑、空间具有了“事件性”，从而上升到四维印象。C．罗厄（和F．寇特尔）引用了噶塞特（Jose Ortega Y Gasset）的一段话来表明历史对于人类的重要性：“人，一言以蔽之，是没有自然的，他所拥有的是……历史。换言之，自然是物，人是历史。猩猩与人的区别不在于理智的有无，而在于猩猩是没有记忆的。这种可怜的生物每天早晨所要面临的都是昨天生活的遗忘……割裂与过去的延续性，就是贬低人的价值和对猩猩的模仿。”任何一类文化中，历史感的把握是人们最重要的体验之一，个体的思考传达其存在感及历史观，个体之思又有赖于知识和文化意识的传递，这就使得不管承认与否，我们事实上生活在传统氛围之中。而特定地域、特定民族的历史意识在文化创造中是无法抹杀和替代的，一方面，“所有历史都是当代史，”另一方面，“五官感觉的形成是以往全部历史的结果”。之所以哈佛、斯坦福、剑桥等名校在人们心中占有一席之地，不仅是由于师资力量的先进，更重要的一点是因为她们承载着记忆，是经过先后一二百年的建设，通过逐渐积累建筑文脉和历史，才逐步形成的有历史底蕴的校园。我们看到的不单是建筑的美，而是她的“场景”，一种透射出百年历史和文化积淀的场景。历史是不能捏造的，追本溯源地看待这些高等学府的校园，设计者在建造之初也一定是从人本关怀的规划设计思想出发，用低调、内敛、有品位的文化建筑来营造生动的校园。这种低调质朴的风格，以及近人怡心的人性尺度空间对于大学校园环境的营造具有积极的借鉴意义。

台湾学者龙应台认为：从传统走向现代的过程中，所有城市都努力往“现代化”的方向走去，但城市发展的目标其实不应该是在“现代”那一端，而是在传统与现代这两点中间。而水产大学的建设也在把握一个历史和文化的中点，一面是传统、一面是现代。

自古以来，建筑师们由于对传统的态度取舍不同，形成了各种流派各种风格。就现代主义和历史主义而言，现代主义忽视对传统精神文化内涵的表达，历史主义忽视对现代物质文化内涵的表达，而我们在水产大学所要做的，是在更广阔的历史和民族文化的背景上，以深邃的眼光来审视中国文化的深层积淀，从中发现和汲取在今天仍然具有生命力的建筑哲学、人文思想、设计意念和形式语言，同新时代、新概念、新的审美需要结合起来，并对某些形式进行适度地抽象，简化和变形，赋予建筑一种公众大多可以认知和解读的符号系统，从而创造出一些有鲜明个性特征和文化内涵的现代建筑形象。中国建筑的发展没有经历哥特时代、文艺复兴、巴洛克、洛可可，也没有新风格运动，历史的断层造就了今天的百花齐放、百家争鸣，轰轰烈烈的城市运动，成就了一大批不同风格的建筑，精品是有的，但浮华的东西更多，水产大学的设计中如何注重传统精神的传承，是值得我们思考的问题。

历史可以借鉴但不能模仿，为什么同样是东方国家的日本，却能够把民族精神发挥到极致，原因在于日本建筑没有一成不变地照搬照抄，而是按照自己特有的文化、审美观念加以融合，从而形成了独特的建筑艺术。从它表现出的平淡、单纯和空灵中，可以感受到所传递出的日本特有的禅宗思想和自然本位的观念。那么，对于水产大学的设计而言，应该深刻理解其设计背景，不仅要从中国传统建筑中吸取文化精神，也要审时度势，使设计理念有现实意义以及可操作的依据。在校区规划和校区建筑设计中既要创新，又要延续，不作封闭的重复，使建筑增加新的意义。因此，它应该是一个文脉和文化传统延续的问题，是将老校区的情感移植到新校区的问题。国外不乏高校建设的成功例子，值得我们借鉴学习，但一种富有生命力的东西必须是在学习别人先进的思想和观念的同时，还要对本土的文化传统作深入的研究。

上海水产大学基地选址位于远离上海市区的南汇洋山港后方基地的临港新城。依托洋山国际深水港、在芦潮港一片广袤滩涂上新建的临港新城，追求的是"生态城市、风景城市、旅游城市、数字城市"的建设目标，学校的入驻将增加新城的人气和人文气息。因为基地位置处于上海市区的周边地区，因此不能仅从上海老建筑中吸取灵感。从大的地域背景下，上海属于中国江南地区，江南地区尤以江南水乡以及江南园林的建筑风格深入人心，小桥流水的生活场景和亲切宜人的建筑尺度，以及以黑白灰为主的谦和的建筑色调，无不体现了江南地区的建筑特色；从小的地域背景下，上海水产大学虽然远离市区，但不能否认的是上海殖民风格建筑印证了上海历史的发展与变迁，水产大学本身不能脱离上海已有的文化底蕴和历史背景；从本身的使用性质而言，校园建筑作为文化建筑的一种，又具备了教育的功能，因此我们可以追求中国传统书院传达的意境，以作为对书院历史与传统文化的隐喻；从时间性看，这无疑是一个现代化的高校，无论是在使用功能、材料使用等方面都应该顺应时代的发展。因此，我们要在设计中探索传统建筑逻辑和现代建筑逻辑、传统审美意识与现代审美意识的结合方式，以达到对上海水产大学建筑群的成功塑造。

中国的书院已存千年，才子们都是在"庭院深深深几许"的环境中进士及第，金榜题名。而传统四合院是中国书院建筑的固有模式，贝聿铭也说："中国传统建筑的精髓是幽深的庭院。"纵观古代书院的发展史，书院建筑的文化品性体现在以下几点：1、朴素的建筑观；2、寄情山水的环境观；3、追求意境的人文观；4、书院建筑空间组织的礼乐精神。表现在选址上，重视历史古迹名人遗迹等人文因素，强调人与自然环境的统一，以达"天人合一"的境界；表现在建筑群落的组织上，充分体现了中国"礼乐"精神之表达；表现在院落空间的塑造上，就像写文章一样，"文如看山不喜平"，随着角度的变化，层层深入，含蓄地表露她的魅力，多用天井穿插、屏风隐蔽，以形成丰富多样的空间层次，给人以宁静之感；表现在建筑色彩上，多采用黑、白、棕等中性色，色感沉着，即可达到强烈的对比，又可获得轻松的调和感。因此，在上海水产大学新校区的规划建筑设计中，我们对大学的场景形式作了自己的诠释：用群落式的建筑布局，递进合院式的空间结构来创作一种能承载中国文化底蕴和西方教育模式的书院式的场景。

水产大学作为中国的现代校园，要在中国大学造园运动中独具特色，应该汲取传统文化的精神，创造出有意味的场所和场所感，正如中国山水画要创造形式之外的意境一样，实现以"远尘世之嚣，聆清幽之胜，踵名贤之迹，兴尚友之思"为意境，以"风声、雨声、读书声、声声入耳，家事、国事、天下事、事事关心"为情趣，不求雕饰华丽，但求宁静清幽。

Experiment Centre of Shanghai Institute of Visual Art, Fudan University

复旦大学视觉艺术学院实验中心

Shanghai, China, 2005.

复旦大学上海视觉艺术学院位于上海市松江区文翔路2200号，是上海松江大学园区的中心地块，紧邻资源共享区，占地748亩。复旦大学上海视觉艺术学院下设艺术设计学院、传媒影视学院、空间与工业设计学院、时尚设计学院和美术学院，已开设艺术设计、动画、广告、摄影、广播电视编导、服装设计与工程及绘画、雕塑等专业。

建设实验中心是复旦大学上海视觉艺术学院进行教学改革探索的重要举措之一。该中心建成后，将成为上海市应用型人才培养模式改革的实验平台，同时也是一个具有国际一流水准的产、学、研结合的实践平台和全面开放的共享平台，并将填补国内大型集合型艺术实验工作室项目的空白。

实验中心位于校区东北角，北靠城市主干道文汇路，南临校园主环道，西面为二期建筑预留用地。由艺术工作室和美术学院用房两部分组成，集教学、研究和制作功能为一体。设有玻璃、陶瓷、木质、石质、金属、绘画、版画、综合材料、多媒体、素描等工作室以及美术教室等，此外还设有多个专家工作室。建筑地上四层，地下一层，四层为美术学院用房，其余各层为各类艺术工作室。

建筑的本体性格决定了设计过程的艰巨性，该艺术实验中心的设计过程中体现了建筑与艺术和技术的艰难融合。美不仅是结果，更重要的是接近美的历程。十多轮的设计案例，跨越了从折衷到现代的多种风格，体现了建筑师与艺术家思想的交流与碰撞。

01 朝来暮去，视觉艺术实验中心是一个各种艺术行为发生的场所。
From morning till night, Experiment Center of SIVA is a place where various artistic activities happen.

02 作为艺术的发生容器，朴素、古典、含蓄、内敛，也许是最合适的。

As a container that creates art, simplicity, classicality, connotation and humility are the most suitable features of it.

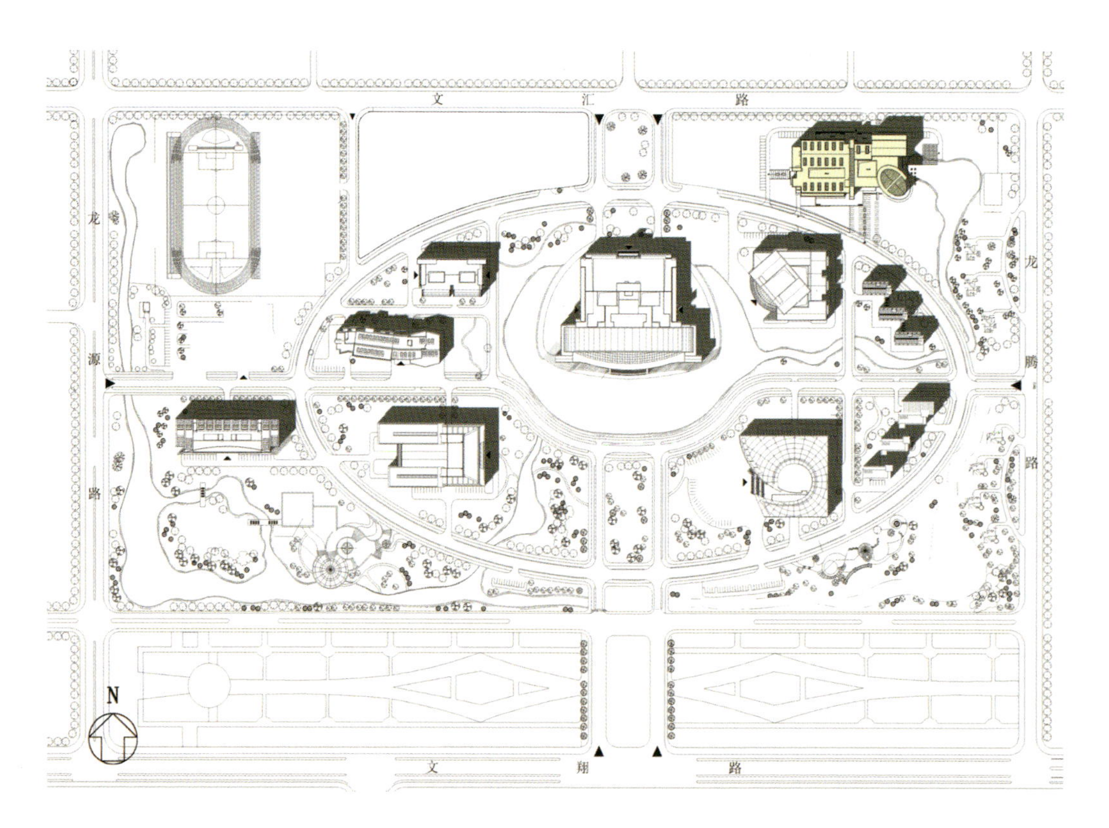

Site plan / 总平面

Located at No. 2200 Wenxiang Road, Songjiang District, Shanghai Institute of Visual Art (SIVA) lies in the center block of Shanghai Songjiang University Town with an area of 490,000 square meters, next to the resource-sharing area. SIVA comprises College of Art Design, College of Media and Films, College of Spatial and Industrial Design, College of Fashion Design and College of Fine Arts. The courses including art design, cartoon, advertising, photography, broadcasting and TV editing, fashion design, painting, sculpture etc.

The construction of the Experiment Center is one of the important steps of SIVA's education reform. When finished, it will become an experimental platform for cultivating mode reform of application-oriented personnel in Shanghai. As a world first-class practice platform for production, study and research and a complete open platform for share, the experiment center will also fill up the demands of comprehensive art studios in China.

The experiment center lies in the north-east part of the campus, north to Wenhui Road, south to ring road of the university and west to the land reserved for second-phase constructions. Consisting of art studios and College of Fine Arts, the experiment center is integrated with education, research and facture. There are numbers of studios and classrooms including glass studio, china studio, wood studio, stone studio, metal studio, drawing studio, print studio, comprehensive material studio, media studio, sketch studio, several expert studios and fine arts classrooms. The building is made up of four levels on the ground and one level underground. The fourth level belongs to College of Fine Arts while others serve for different studios.

The ontological feature of buildings makes the design process difficult and hard. The whole design process of experiment center embodies the difficult integration of architecture, art and technology. Beauty is not only a result but more important of a process approaching itself. With more than ten rounds of design modifications and the leap from neoclassic to modern style, all of these reflect the conversation and conflict between architects and artists.

基本资料		BASIC INFORMATION	
地理位置：	上海市松江区文翔路 2200 号	Location:	Wenxiang Road, Songjiang District, Shanghai
用地面积：	24,023 平方米	Base area:	24,023m^2
建筑面积：	21,025平方米	Building area:	21,025m^2
占地面积：	6,109平方米	Site area:	6,109m^2
容积率：	0.63	Plot ratio:	0.63
绿化率：	39.6％	Green ratio:	39.6 %
设计时间：	2005年	Time:	2005

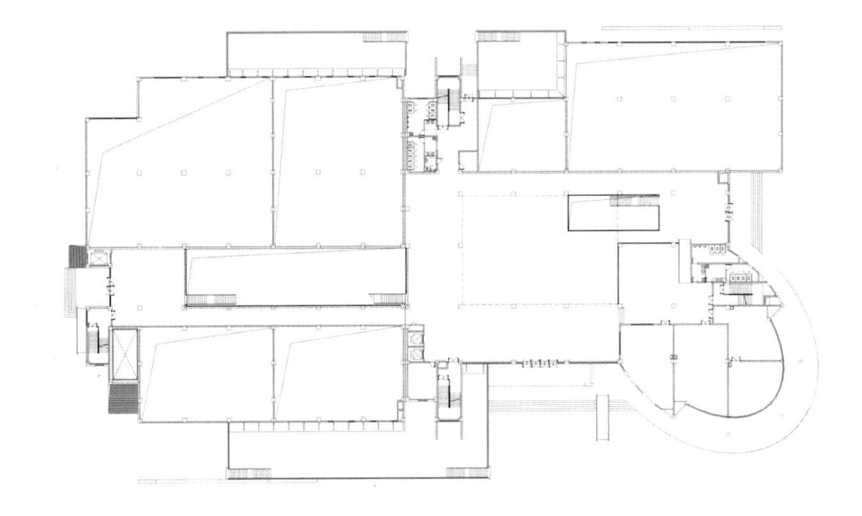

Fisrt floor / 一层平面

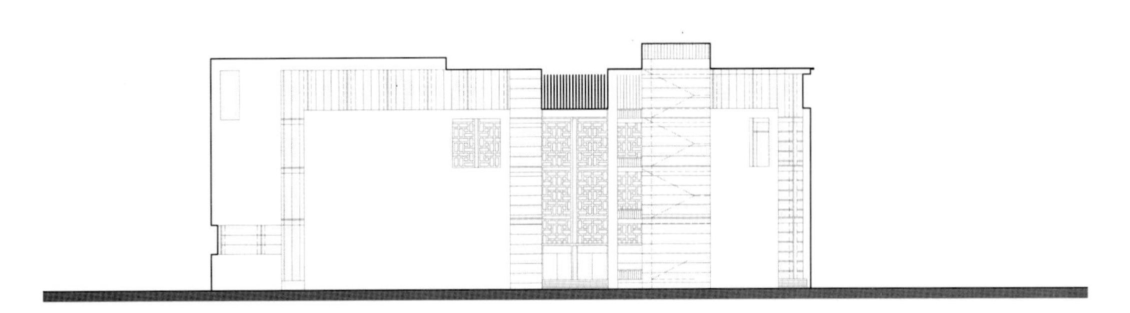

facade / 立面

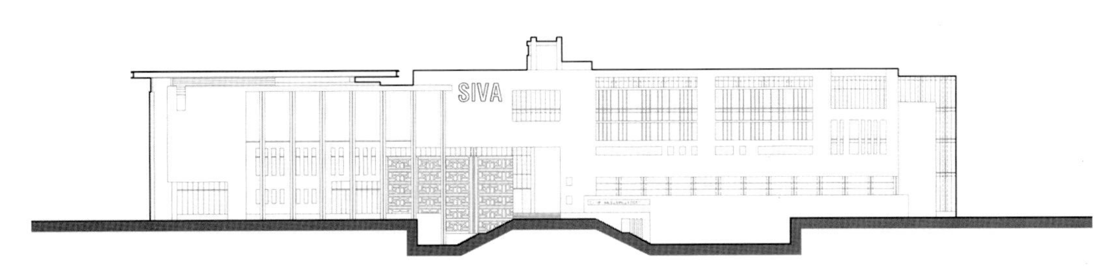

facade / 立面

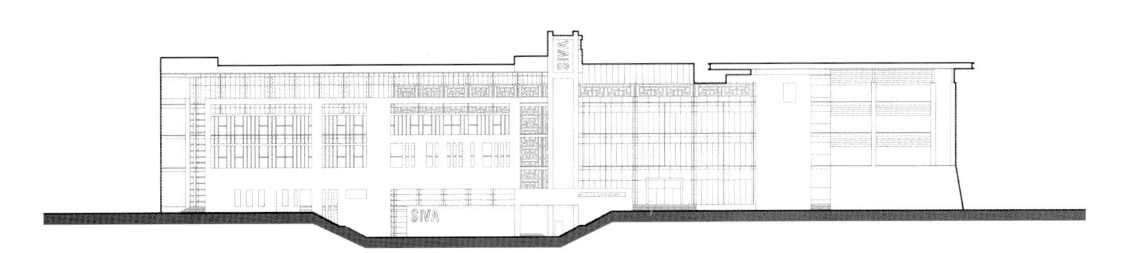

facade / 立面

项目名称	建筑面积(㎡)	序号	项目名称	建筑面积(㎡)
食堂	18000	19	学生宿舍	150000
学生食堂(一期)	8000		19-1\19-2(一期多层)	8000
学生食堂(二期)	10000		19-3\19-4\19-5(一期小高层)	36000
交流中心(三期)	15000		19-6\——19-12(二期多层)	70000
教师公寓(二期)	6500		19-13——19-15(二期小高层)	36000
设施配套用房	4700	20	400米标准田径场(一期)(三期)	2
：车队绿化维修(二期)	500	21	游泳馆(三期)	5000
(一期)	700	22	篮\排\网球场(一期)(二期)(三期)	60
变电站(一期)	500	23	预留用地	90143
(二期)	3000		总建筑面积	466850
综合用房(二期)	4000		其中，一期建筑面积	143200
用房(二期)	1166		二期建筑面积	255279
政用房(二期)	9000		三期建筑面积	68371

02 新校区总体详细规划总平面图。
Detailed site plan.

03 校区总体规划动画场景组图，充分展示校园丰富的建筑景观空间设计构想。
Movie clips of SIT, display the concept of architectural and landscape design.

A 公共教学楼

用地面积：	29,079m^2
建筑面积：	24,120m^2
占地面积：	7,491m^2
建筑密度：	25.76 %
容积率：	0.83
绿化率：	26.1%

B 计算机中心

用地面积：	18,350m^2
建筑面积：	10,077m^2
占地面积：	2,722m^2
建筑密度：	14.8 %
容积率：	0.55
绿化率：	43 %

C 数理楼

用地面积：	9,750m^2
占地面积：	2,112m^2
建筑面积：	7,440m^2
其中：	
地上建筑面积：	7,324m^2
地下建筑面积：	115m^2
建筑密度：	22.69%
容积率：	0.76
绿化率：	35.7%

D 图文信息中心

用地面积：	36,930m^2
占地面积:	6,700m^2
建筑面积：	30,627m^2
其中：	
地上建筑面积：	27,319m^2
地下建筑面积：	3,308m^2
建筑密度：	18.1%
容积率：	0.74
绿地率：	35.6%

A Public Teaching Building

Base area:	29,079m^2
Building area:	24,120m^2
Site area:	7,491m^2
Building density:	25.76 %
Plot ratio:	0.83
Green ratio:	26.1%

B Computer Centre

Base area:	18,350m^2
Building area:	10,077m^2
Site area:	2,722m^2
Building density:	14.8 %
Plot ratio:	0.55
Green ratio:	43 %

C Science Building

Base area:	9,750m^2
Building area:	2,112m^2
Site area:	7,440m^2
Including:	
On the ground area:	7,324m^2
Underground area:	115m^2
Building density:	22.69%
Plot ratio:	0.76
Green ratio:	35.7%

D Information Centre

Base area:	36,930m^2
Building area:	6,700m^2
Site area:	30,627m^2
Including:	
On the ground area:	27,319m^2
Underground area:	3,308m^2
Building density:	18.1%
Plot ratio:	0.74
Green ratio:	35.6%

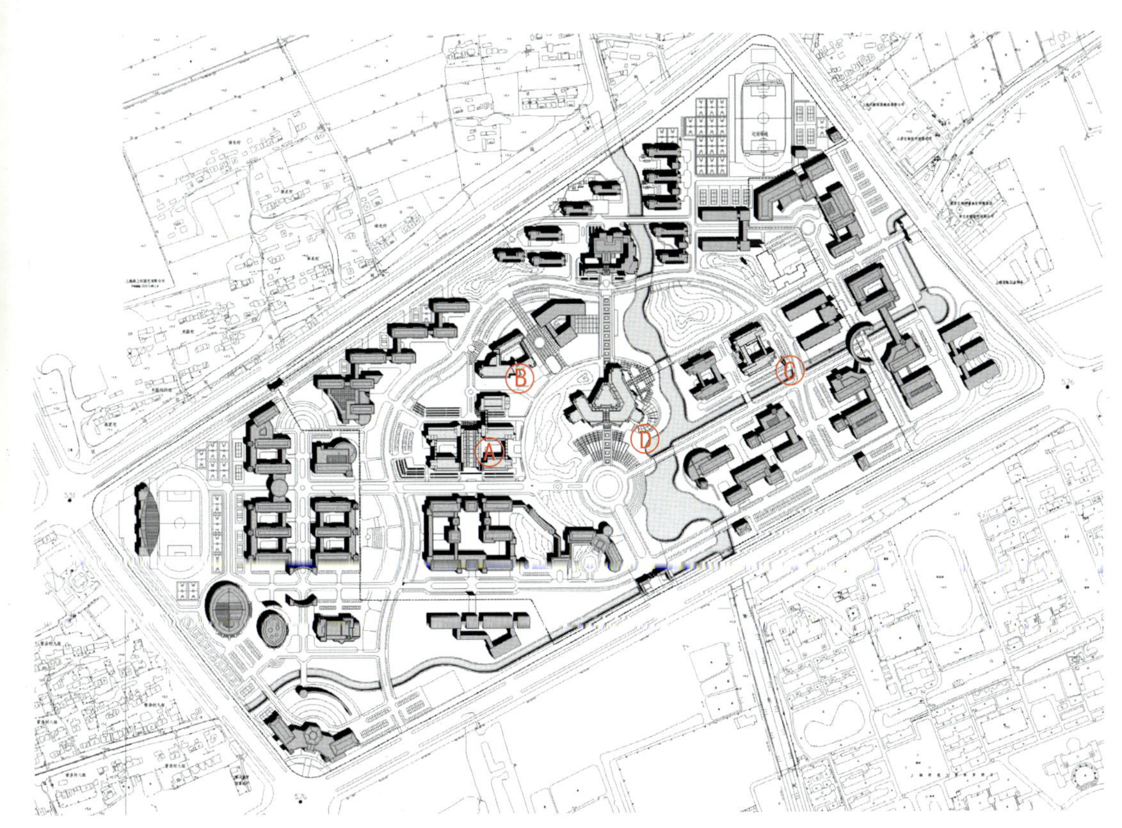

Site plan / 总平面

公共教学楼

公共教学楼南临主干道，东临校园中央景观轴，由五个教学单元组成。240 人大教室集中设置在基地北部（三层），其余四个教学单元（四层）组合成庭院空间。单元之间以廊道相连，便于师生交流。教学单元中央围合成的内院为师生们提供了交流、学习、休憩等多种类、多层次的交往空间。建筑外挑两层高的门廊丰富了入口的空间层次，强化了建筑形象。

Public Teaching Building

To the south of the public teaching building is the main highway, to the east is the central view stalk in the campus. The building is composed by five teaching units. The big classroom which can hold 240 students is established to the north of the base(3 floors) and other four teaching units(4 layer) are assemble as a space of yard. Units are connected with each other with gallery so that it is convenient to the communication of teachers and students. The internal yard which is surrounded by teaching units offers a multi-sort and multi-level communication space for teachers and students to communicate with each other, study and relax. The entrance hall whose height is two floors enrich the space arrangement of the entrance outside the building and emphasis the building image.

04 入口空间的大门厅采用景观建筑的设计手法，将人流引入内院再分向各个单体。

The entrance hall is designed by landscape architecture techniques. People are first led to the inside and then separated to different buildings.

05 教学楼立面采用古典横三纵五的构图方法。

The facade of the public teaching building is designed by the composition that contains three in plane and five in vertical.

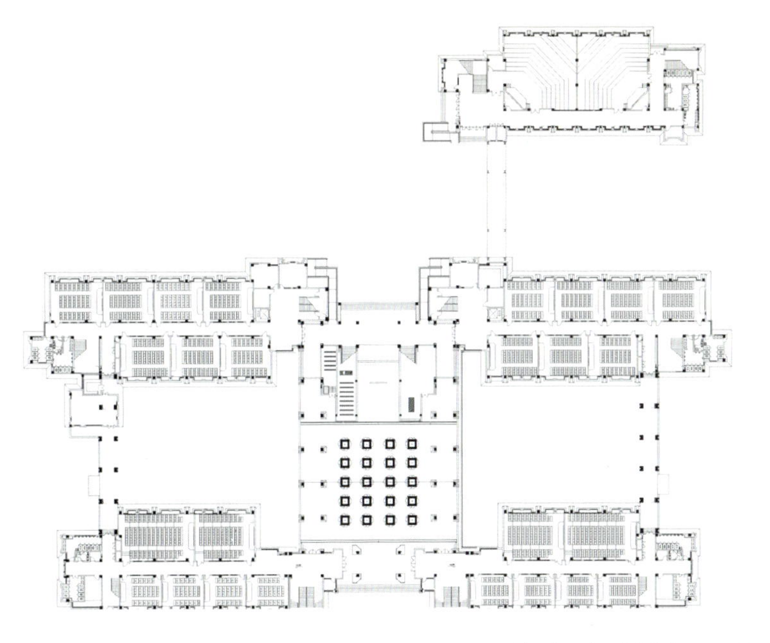

First floor / 公共教学楼一层平面图

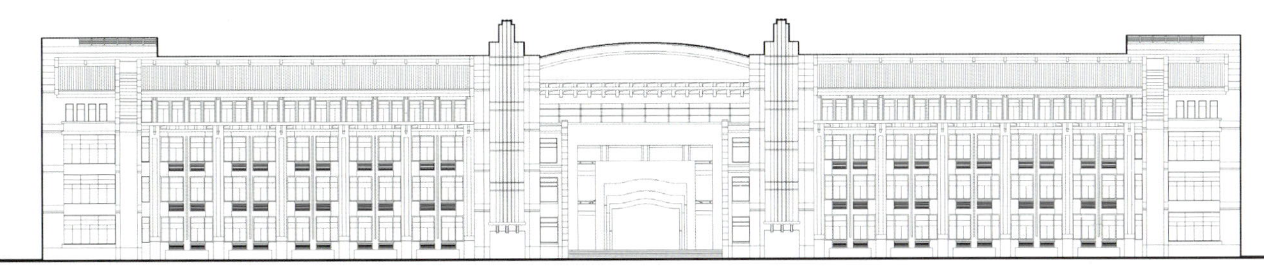

South facade / 公共教学楼南立面

计算机中心

计算机中心北临校园规划环路，南面与教学楼遥相呼应，东面为预留建筑用地。建筑共四层，包括标准机房、综合机房和大机房及办公空间。计算中心成四合院制式，内部形成庭院。庭院朝向校园的虚轴，并与对面的教学楼、图文信息楼遥相呼应。建筑环境优美，功能合理，流线明晰。

Computer Centre

The north of the computer center faces the campus programming wreath road, south to the teaching building, east to the reserve building site. The building is 4 layers, including standard computer room, synthesizing computer room, large computer room and office space. The computer center is a type of the quadrangle. The inner part of it forms a courtyard. The courtyard faces the falsely stalk of the campus and faces the opposite teaching building and the diagram and text information building as well. The environment of the building is beautiful; the function is reasonable and the flowing line is clear.

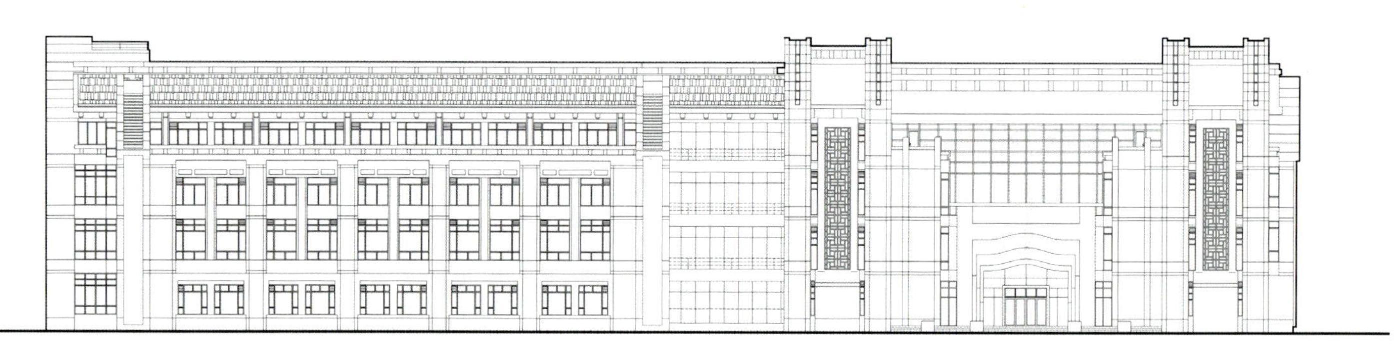

North facade / 计算机中心北立面

06 计算机中心鸟瞰。
Airscape of Computer Centre.

07 计算机中心巧妙结合用地，在角部设计了半圆形的入口门厅。
We took good advantage of the site in design of the computer centre. A semicircle entrance hall is added on the corner.

08 计算机中心次入口形如放大的门阙，有浓郁的中国传统建筑的味道。
The shape of the sub-entrance of the computer centre is like an enlarged 'Men-que', It has a strong taste of Chinese tradition.

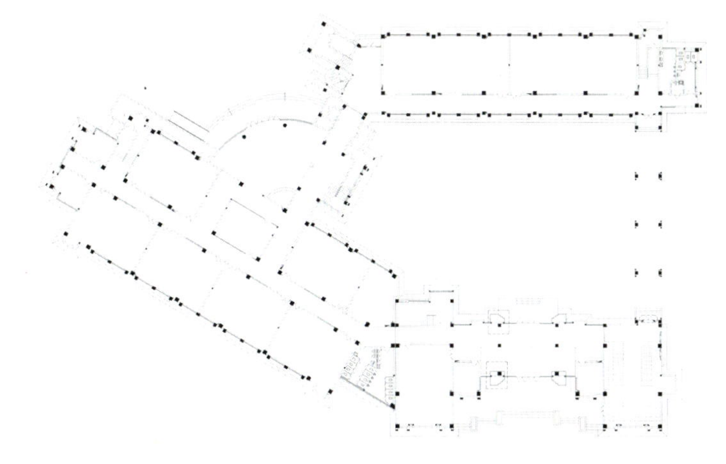

Fisrt floor / 计算机中心一层平面

07

08

数理楼

数理楼南面临水，北临规划中的校园主要人行道路，东临规划中的校园环路，西面与校园预留绿化用地及图文信息中心遥相呼应。主要由实验用房和行政办公用房组成，实验用房主要分布在一层和北楼，行政办公用房主要集中在二至四层的南楼。

主入口广场西向设置，可与二期建设的经管、人文、社科外语学科楼形成院系活动场所；南北两楼之间形成庭院，庭院透过架空的连廊朝向校园主干道，并与对面的工程训练中心遥相呼应。它不但成为联系南北两楼的核心枢纽，还能兼用于学生课外休闲娱乐活动。

Science Building

To the west and to the south of the mathematics building faces water, to the north faces the main sidewalk road of the campus programming, to the south faces the campus programming wreath road, to the west faces the diagram and text information center, to the east faces the reserve building site. The building is mainly composed by the laboratories and the administration offices. The laboratories are put on the first floor and in the north building and the administration offices are mainly put on the second floor to the 4th floor in the south building.

Between the north building and the south building is a courtyard. The courtyard faces the sinuous river surface and faces the opposite diagram and text information center as well. Not only has it become the core vital point which contacts the south building and the north building, but also used for students to review the lessons and the result exhibit.

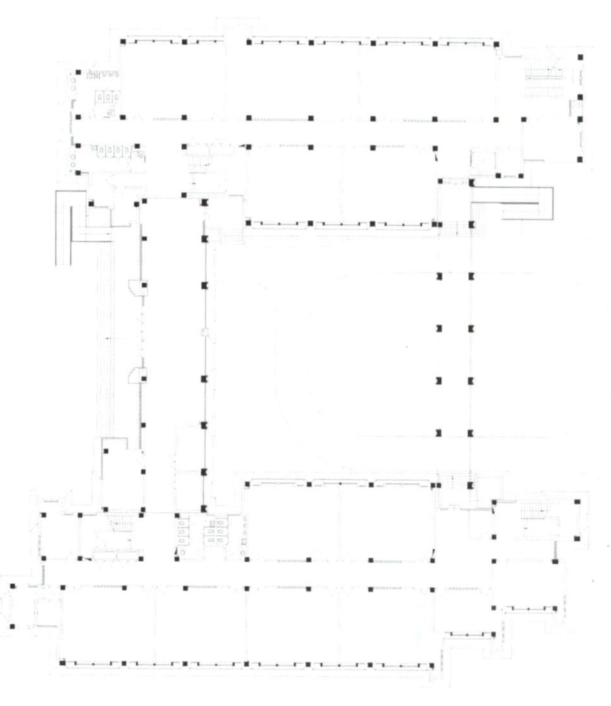

Fisrt floor / 数理楼一层平面

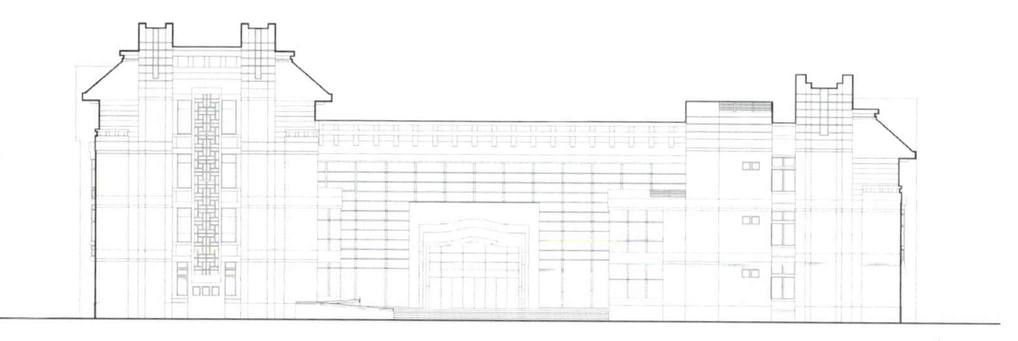

West facade / 数理楼西立面

09 数理楼的墙身抽象自中式传统建筑。
The designing concept of the wall is coming from Chinese traditional buildings.

10 入口门头，层层叠涩。
The doorframe, with many layers.

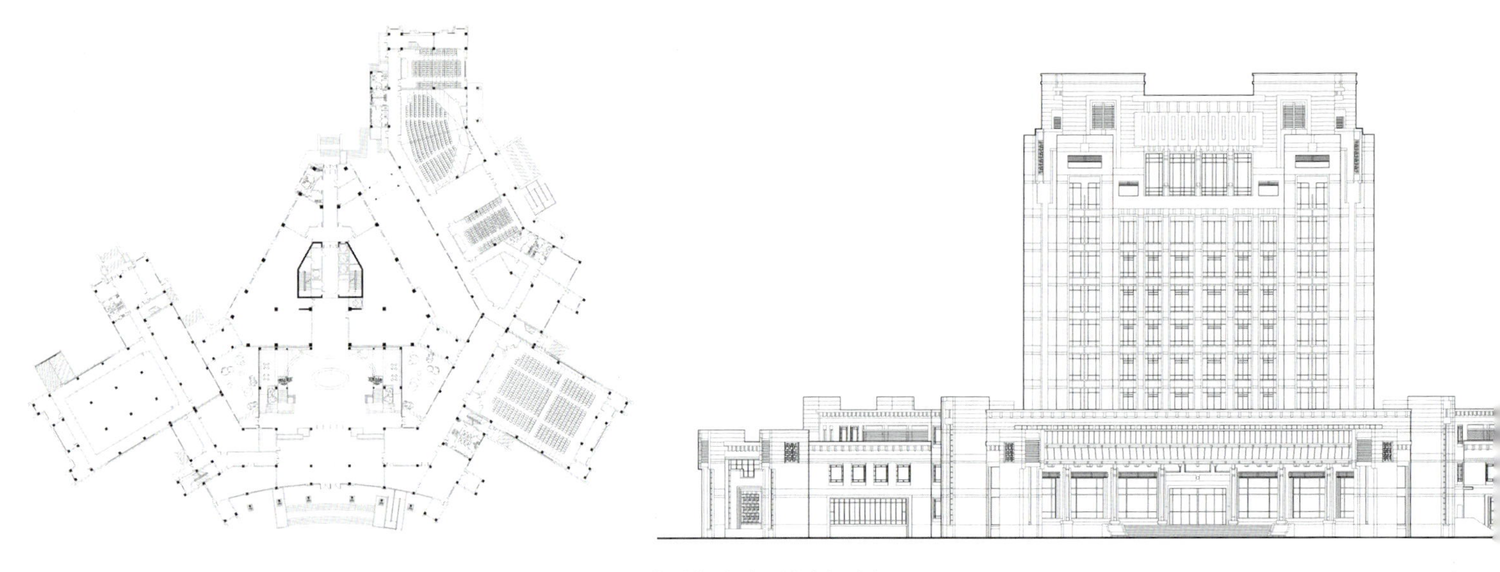

Fisrt floor / 图文信息中心一层平面

South facade / 图文信息中心南立面

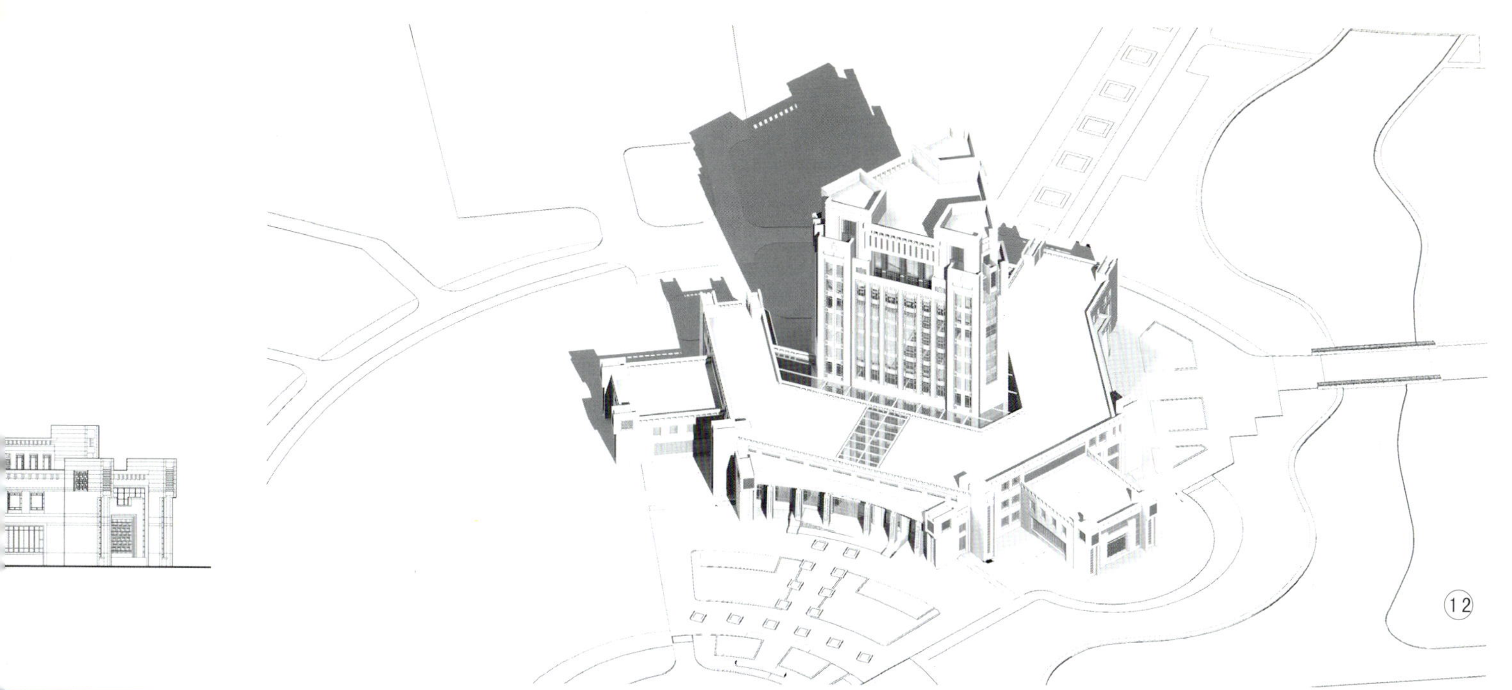

图文信息中心

图文信息中心南临校园中央景观轴，与校园景观轴呈一定偏角。由图书馆、会议中心、档案馆、网络中心四部分组成。由于会议中心的独立性，将其设置裙房之中 ，设有独立出入口，与图书馆之间以绿化庭院隔开。网络中心由于设备比较重，设置在一层。档案馆则设置在二层。图书馆设置在塔楼内，二三层为借阅处，四层为中文及期刊阅览室，五层设置电子阅览室，六层为外文期刊室，七层外文阅览室，八层为文献检索，九层为艺术阅览室，顶层为校史陈列观光层。图文信息中心塔楼形式为三角形，体现了对规划各向关系的呼应。

Information Centre

The south of the information center faces the central view stalk of the campus, presenting a certain angle with the campus view stalk. The center is composed by four parts which are the library, the meeting center, the file building, and the network center. Because of the independence of the meeting center, it is established in the skirt building. The center is established with an independent entrance and is separated with library by green zone. The network center is put on the first floor because the equipments is heavy. The file building is put on the second floor. The library is put in the tower. On the second floor and the third floor is a place for borrowing and reading; on the 4th floor is a Chinese book and magazine reading room; on the 5th floor is a electronic reading room; on the 6th floor is the foreign magazine reading room; on the 7th floor is the foreign book reading room; on the 8th floor is the literature search; on the 9th floor is the art reading room and the top floor is a visiting floor exhibiting the school history. The form of the tower is a triangle, showing that each relative of programming acts in cooperation.

11 图书馆选用三角形平面，从校园各个方位都有很好的观瞻效果。

The library is a triangle in plane arrangement. Seen from different directions on campus, It has a good view effect.

12 图书馆的鸟瞰三维模型。

Airscape of 3D model.

大学印象

——上海应用技术学院奉贤校区规划设计

我们这一代人说起大学，让人不由地想起《校园民谣》中的一张黑白背景画——照片中一长排破旧的自行车站在初冬的法国梧桐树旁，树枝上一群群麻雀探头探脑，忽然被下课的铃声惊得飞向青天。大学给我们的感觉总是书生意气的——背着书包走在秋天的林荫道上，满地金黄的梧桐树叶，嘎吱嘎吱的响声在脚底散开。春天的早晨也是很美丽的，稀薄的阳光散布在空气里，植物在茁壮地疯狂地生长，带着一颗充满希望的心站在阳光下，光晕洒在浓荫密布的小红楼的屋檐上，那是一种非常奇妙的感觉。

大学，在人们心中应该是一片纯净的绿地吧！没有世俗的流尘，亦没有欲望的漩涡。高等学府总被人施以象牙塔、神圣殿堂等等纯净之地的称谓，我们在实践上海应用技术学院奉贤校区规划的过程中，思考大学的真正意义的旅途上，记录了设计师对大学的印象，从中寄语上海应用技术学院奉贤校区的整体设计。

双面绣

从传统走向现代的过程中，所有城市都努力往“现代化”的方向走去，但城市发展的目标其实不应该是在“现代”那一端，而是在传统与现代这两点中间。中国的城市发展就如同“双面苏绣”，一面是科技，一面是人文；一面是传统文化，一面是现代经济；协调地合为一体。

上海应用技术学院新校区作为城市的缩影也必须符合城市发展的文脉，顺应时代发展的潮流，一面是传统、一面是现代，恰似一个历史和文化的“双面绣”。

现实世界中新建大学“文化生态”的失落

最近十几年，高校建设突飞猛进，各地出现了许多万人大学乃至数万人的巨型大学。人们经常把大学比做一个小城市，但当一个小城市突然发展成一个大型或巨型城市时，就会产生许多中、小城市未曾遇到的矛盾。

当前的校园建设，人们似乎也喜欢越大越好，以显其气派。国外的许多号称数万学生的大学，大多是分布在各州、市建分校。集中在一个地区的大学校园规模到底多大合适尚无定论。前苏联曾规定合理规模的大型校园为6000~12000名学生；欧洲大陆一般与论认为综合性大学的合理规模是6000名学生；有的则以学生的步行距离5分钟(即400m)为半径来限定校园的规模。

中国人口多，需要建一些大型校园才能满足逐步扩大招生规模的需求，但大型校园在运营中有什么问题，大型或特大型校园在规划上有什么新特点，这些问题都有待于在实践中进行探讨与总结。

清华校园扩展的过程中，限于地段，教学区向南发展，学生生活区往北扩展，学生区与新建的教学区越离越远，最长距离大约1．5km。南北两区靠两条道路相连，呈哑铃状。于是清华园里的一大特点就是自行车比人多(算上丢在一旁不能骑的旧车)，上下课时道路是车的洪流，而教室周围和校园广场都变成了黑压压的停放自行车的海洋。

下课后看不到师生在校园继续切磋，看不到同学三三两两在道路上边走边聊，交流信息，到处则是急匆匆赶路的人，还要时刻注意不被别的车碰倒。一位研究生曾做了一项关于校内骑车的安全感问题的调查，有70%的同学认为不安全。

一位出国多年的老校友，过去是在老教学区生活，现在又见到豪华气派的新区，谈到对新老两区的印象时说："到了老区想念书，到了新区想挣钱"，"新区有点像曼哈顿"。

清华的大多师生在承认新区建筑新颖、气派的同时却更喜欢老区，外来参观者也大多有同感。

老区的建筑多为2~4层，造型典雅、朴素，校区空间尺度宜人，绿荫浓郁，芳草茵茵，环境宁静而温馨，烘托出高等学府文化氛围。在这里，人们自然想的是要求知，要念书。其实"念书"和"挣钱"都是好事，但是，最能体现校园环境本质和灵魂的东西是什么呢？

"人文环境"在大学校园中集中体现在"文化生态"方面。校园可以有"自然生态"、"社会生态"和"文化生态"等多种生态化表现。

"文化生态"不单是表层的刻字、树碑，更多的是隐形的精神境界，校园环境有无促进人们产生求知和探索的欲望？是否促进人们的交往与研讨？是否处处有利于学生身心的培养和素质的健康成长？它对于校园规划肌理提出什么要求？如何在规划中体现这种意境。

大学的场所精神

今天的校园正由原来内向的、封闭的，游离于城市之外的隐士型学院，转变为功能复杂、多层次、多元化的社会综合体。"多才雅得江山助"，人才的培养，不仅在于课堂知识的传授，而人才品格、情操的培养，也很大程度地依赖于校园环境潜移默化的作用。如何在比较独立又狭长的用地中塑造独特的校园氛围，以区别于众多设计思想丰富但确没有个性的公共空间，不仅要满足校区建设的功能性要求，而且要重视高校学府的环境育人作用，寻求其文化氛围中空间的精神本质，即探求其场所精神。

场所精神一词来自于古罗马，根据古罗马人的信仰，每一个独立的本体都有自己的灵魂，这种灵魂赋予人和场所生命，同时也决定他们的特性与本质。在我们一般的概念里，所谓场所或空间结构是一种固定而永久的状态，因为现代环境内外在的单调特性，使得今日的场所个性几乎沦丧。很明显，在中国或世界某些地方，传统的人类聚落的特质已经无可挽回的沦丧了，无论都市纹理、生活空间、建筑、街道……，不断重复着单调贫乏的步伐，人们不同的交往空间已经失去了各自的意义而成为广泛的扩张，更可以说是缺乏特性，可以说，由于场所精神的丧失，人们呈现越来越强烈的失落感与缺乏认同感。

不知道应用技术学院新校区这片土地是不是浮华城市中仅剩的净土之一，当我们怀念起校园生活，怀念她绿树浓郁，芳草荫荫，宁静而温馨，悦情而怡性时，我们不希望她急于流入各种事物扩张的洪流中而失去她原有的真实，不愿意看到认同感的变异造成对其本质的印象被疏离感所取代。我们应该在校区的空间塑造上体现更多的人文关怀，不仅能使校区无愧为社会综合体，更能让我们一边漫步在校园里一边能够亲切地回忆起她的本质。

应用技术学院奉贤校区的"五重奏"

一、非线性的空间构图手法

传统大学校园追求的线性对称式空间构图手法已落在我们时代巨人之后，我们在应用技术学院新校区的规划设计上采用非线性的整体空间构图模式，主轴的连动与转折；校园肌理的变相与呼应；在教学建筑空间处理上引入了校园的街景概念；偏转的图书馆等都是我们在复旦大学和华东政法学院建筑设计上的质的突破，意图从根本上打破传统校园的对称式的构图模式。

二、教学与生活尺度回归

拉近生活区与学习区的距离，避免教学区域"人去楼空"的悲凉（从教育理念的角度调整校区布局以及营造过渡性场所)。很多校区规划都是从一片完整的土地上开始，规划中功能分区结构明确，生活区与学习区泾渭分明。由于学校扩招，很多学校每到上学放学时间教学区域总是人头攒动。下课之余大家都在匆匆的赶路途中，校园内顿时交通混乱，失去了一些能让人驻足停留可以安静思考的场所。人去楼空的教学区顿时变得空寂灰暗，没有了活力，因此良好的规划应该为学生的学习与生活提供安全便捷的途径，以便营造出三五成群、边走边聊的人性化和生活化的街道空间。另外，在生活区和教学区的有限距离范围之内，应该提供多处可驻足停留的场所，使学生在学习之余可以休息、交流、感怀沉思。

三、呼吸校园

为体现可持续发展战略，在满足新校园功能构架的条件下，在规划结构中嵌入“片状”与“带状”预留用地。

内环建筑的紧凑布置既解决了各功能组团之间紧密互动的问题，又为未来的发展在周边预留了大片绿地。这些周边的绿地不仅有利于营造校园的自然环境，给周边的行人留下了绿色大学的良好印象，又为将来的校园扩张预留空间。

在各功能组团与核心区外围留有许多“带状”绿地，由于呈“斑状”，在不影响建筑布局的紧凑性的前提下，使整个校园规划具有透气性，恰似一个会呼吸的校园。

四、建筑形式语言的价值认同

全新的校园环境难免会让师生产生情感上的失落，校园形体环境作为一种符号系统，蕴含着复杂多样的意义，传播着丰富的历史内涵。新校区的建筑造型面临着一个文脉延续的问题，如何把老校区情感移植到新校区，我们从老校区的古典元素中汲取养分，创造专属于应用技术大学的建筑样式。既不是五六十年代的苏式中国大屋顶，也不是具体的希腊式、罗马式；既不是巴洛克、也不是洛可可；既不是歌特式，也不是日式。新应用技术学院的建筑样式可能既会有一些中国式建筑的符号，有一些英格兰建筑的色彩，有一些地中海建筑的质感，有一些西班牙海滩建筑的风情，有一些后工业时代的模块等等，摒弃现代主义标榜的纯粹，追求历史认同与社会认同的建筑折衷美学。

相似色法则——到过英国剑桥大学、牛津大学的人，无不为其浑然天成的建筑经典所折服，其建筑至上而下的色彩变化，是那么自然肯定。我想对上海应用技术学院的整体色调的把握，应该很好地借鉴。熟悉建筑历史的人都知道，中国的历史建筑没有色彩，应用技术学院的相似色法则的运用是一个至少可以改变中国高校单色系建筑色彩现状的重要特征。

打破传统的立体屋面建构手法——我们细心地发现，上海高校的屋顶更多地采用小屋顶，均没有采用大屋顶，且屋顶没有独特的自我个性。我们应用技术学院南校区给了我很多启示，我们在应用技术学院新校区采用大屋顶的立体建构手法，可以更好地彰显校园的独特性格，这是应用技术学院特定的屋顶文化。

五、群形态的整体变异

变异在建筑语言分两种，一种是建筑个体符号的变异，一种是强调建筑群体特征的整体变异。上海华东政法学院和工程技术大学所采用的是建筑个体符号的变异，在整体校园传承了老校的特征后，华东政法的图书馆塔楼和工技大的M形图书馆产生了个体变异。但是，上海应用技术学院由于老校区整体建筑形态和中国大多数普遍意义上的高校趋同，所以我们强调了应用技术学院新校区的建筑群形态的整体变异，这就需要我们摒弃所谓“中国特色”，采用极端的“历史折衷”的手法，确立“应用技术学院新校区”的建筑特色，结合一些中国式建筑的符号，借鉴一些英格兰建筑的色彩特征，追求一些地中海建筑的质感，吸取一些西班牙建筑的场景，创造一些后工业时代的模块，走出高度折衷的“应用技术学院建筑美学”。

大学，这片尘世的净土，它似乎将隐藏的所有的人文气息化成一条河流，缓缓地，平静地流淌着，日复一日，年复一年，静默而单纯。上海应用技术学院在这个以经济、商业为象征的大都市中，用她对于学术的严谨与内敛在上海的奉贤海湾营造出了一份深厚的文化底蕴与内涵，虽然不像南京路那样繁华，不似外滩那般美伦美奂，也没有城隍庙的喧嚣，但它就是它，与这座开足引擎的城市一样，在迅速的成长过程中保留下许多美好的历史痕迹，为她自己赢得那值得永恒铭记的百年一瞬……

著名建筑师小沙里宁曾说："大学是属于我们的时代，而庙宇、教堂是中世纪的产物。大学就象我们时代文化沙漠中的绿洲，校园里有美好的步行区、林荫道，而别处都没有。这就说明只有校园才存在永恒的建筑学。"大学校园，已经成为人们心目中理想生活环境的一种象征：绿色、安宁、自由、开放……成为大学校园规划设计所追求的理想境界。

在张爱玲的小说中，老上海是颓美的。斗转星移，光阴荏苒，在20世纪60年代的上海滩，这里有老式弄堂的十里洋场，这里有亘古的流风余韵，这里有彩格的欧式玻璃窗，牧师的颂词，唱诗班的动听歌声；欧洲建筑令人应接不暇，外滩用她那特有的那些古典欧式建筑隐隐约约地告诉着我们——她曾有过的风情万种、缠绵悱恻。黄浦江的海风始终是温柔的，远眺上海在历史进程中留下的印记，深深沉思那长长的历史轨迹，蓦地发现，只要心中藏有永不枯竭的源泉，那前进的道路就永无尽头……这种古老的气息，也似乎让我们看到了上海应用技术学院美好的未来。

教学楼鸟瞰图

第三章

华谏建筑过去时

Chapter 3

DHGP in China building

Science & Research Building and Sports Centre, Shanghai Maritime University

上海海事大学临港新校区科研楼、体育中心

Shanghai, China, 2004

上海海事大学是一所以航运技术、经济与管理为特色的具有工学、管理学、经济学、法学和文学等学科门类的多科性大学。上海海事大学临港新校区位于上海市临港新城内，总规划用地面积约133万m^2。

科研楼

位于校区西南侧主入口左侧，紧临教学楼群。建筑用地面积为10,975m^2。建筑分科研楼和研究生院两部分。科研楼共四层，功能以科研、实验及办公为主；研究生院共四层，功能以办公，教室及会议为主。

研究生院与科研楼之间有着紧密的联系，两者的使用者间良好的相互交流可以促进各自的发展，而之间的庭院就形成最好的交流场所。建筑的错落和场地的不同标高使得庭院空间丰富，并且在庭院中又套有一个圆形下沉庭院，周遭水面平静，绿树环绕，学子们沿阶而坐，或捧卷静读，或掩卷沉思，或两三人低头耳语，与校园气氛融合。

建筑立面现代，色彩素雅，在西面和南面大量的使用了百叶，既可以遮阳，又可以巧妙隐藏空调室外机。

体育中心

位于校区南段，新城规划B1道路与芦潮引河交叉口的北侧，西南临教学楼群，总建筑面积约24,000m^2，包含体育馆、游泳馆、训练馆、田径场四个部分。

总体布局上体育馆、游泳馆、田径场三者面向校园核心区自然围合出一个弧形界面，共同围合出一个开阔的椭圆形广场，满足校园大型文艺演出、集会的需求，同时能够强调出体育中心的群体性格。

建筑层面上体育馆、游泳馆、训练馆三个相对独立的个体形成一个互成角度的向心连接，三馆交合处自然形成一个半开敞半围合的灰空间，表达体育中心的巨大张力。建筑个体的造型手法以“源之点滴，瀚于海洋”为意念，将体育馆、游泳馆、训练馆取意大小不一的三滴水，通过一个整体的建筑表皮处理手法，形成三滴水正欲融合姿态。更表达着一个更为哲理性的构想：渺小水滴经过无数次融合，最终能够汇成浩瀚无际的海洋 。

这两个案例是本书作品中为数不多的没有实现的项目，但其本身所表现出来的建筑美学价值有一定的可取之处，虽是有不少遗憾，但也将其列入书中，以表慰籍。

01 科研楼的设计通过对建筑内部功能合理组织，营造出一个可供师生交流的半开敞图形内庭。

The design constructs a half-open round courtyard for communication between students and teachers by properly organizing the interior functions of the building.

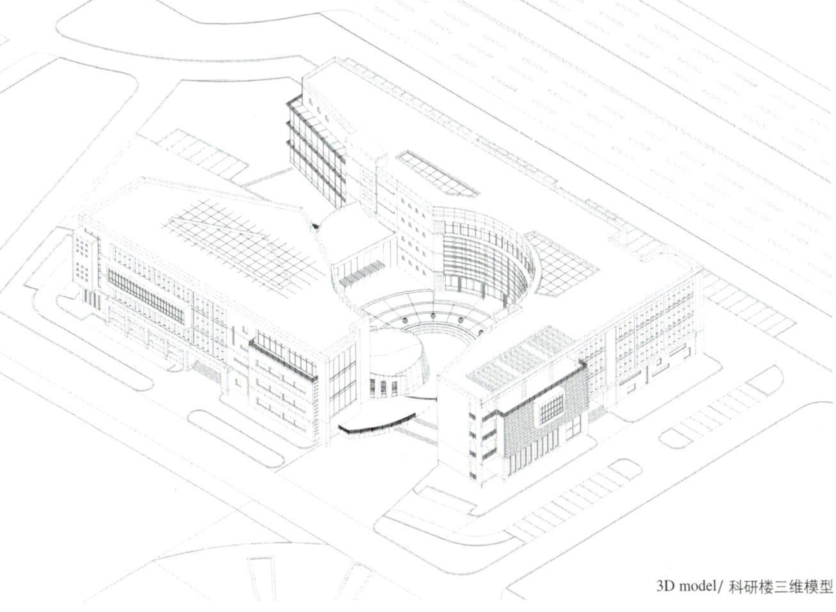

3D model/ 科研楼三维模型

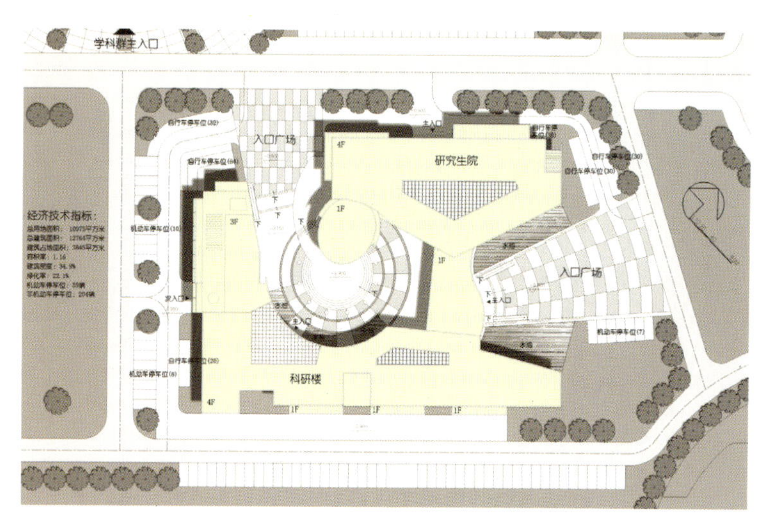

Site plan / 总平面

基本资料：

地理位置：	上海市南汇区临港新城
校区总用地面积	1,330,000m²
科研楼	
总用地面积：	10,975m²
总建筑面积：	12,764m²
其中	
科研楼：	7,428m²
研究生院：	5,156m²
建筑占地面积：	3,845m²
容积率：	1.16
绿化率：	22.1%
设计时间：	2004年

BASIC INFORMATION

Location:	Nanhui District, Shanghai
Total Base area:	1,330,000m²
Science&Research Building	
Total Base area:	10,975m²
Total Building area:	12,764m²
Including	
scientific research center:	7,428m²
graduate school:	5,156m²
Site area:	3,845m²
Plot ratio:	1.16
Green ratio:	22.1%
Time:	2004

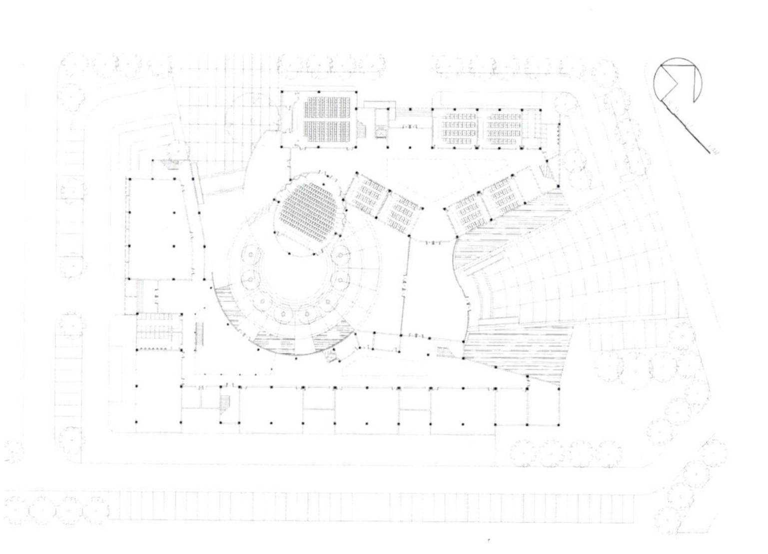

Fisrt floor / 科研楼一层平面

Shanghai Maritime University (SMU) is a multi-discipline university with six fields of study: engineering, management, economics, literature, law and science. SMU Lingang Campus is located in Lingang New Town with a total site area of 1, 330,000 square meters.

Science & Research Building

Science & Research Building is located in the left side of the main entrance in south-west part of the campus, quite near to the teaching complex, which covers an area of 10,975 square meters. Science & Research Building consists of a scientific research center and a graduate school. The four-story scientific research center serves for scientific research, experiment and management. While the four-story graduate school is mainly for teaching, conference and management.

There is a close connection between the scientific research center and the graduate school. The good communication between them can very well complement and promote each other. Thus the courtyard inside becomes the best place for communication. The buildings with different heights and grounds with different levels make the courtyard space interesting and active. There is a round sunken yard inside it with water and trees around it. Students and staffs can sit, read, think and relax.

The facade is decorated with modern style and simple but elegant colors. In south and west elevation, jalousies are used to shade sunlight and hide air-conditions.

02 科研楼用一种包容开放的姿态面向校园主轴，打破传统山墙面的设计手法，用五个相互错位的界面围合出一个入口边庭。
Science & Research building is standing along side the main axis with an open and tolerant attitude. In design, we use five interlaced interfaces in stead of traditional gables to encompass an entrance.

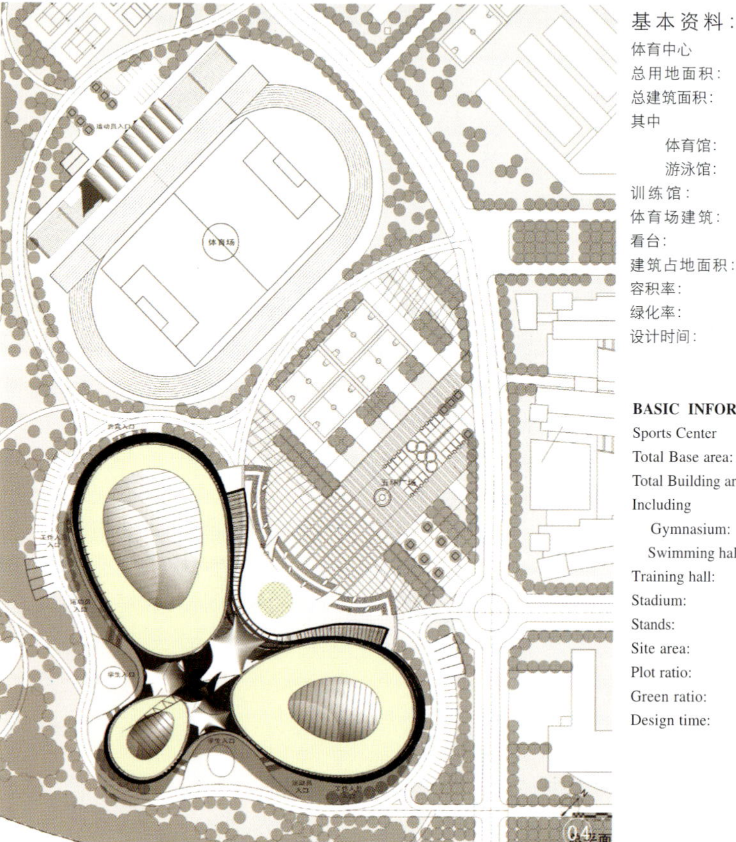

基本资料：

体育中心	
总用地面积：	95,300m²
总建筑面积：	26,250m²
其中	
体育馆：	10,775m²
游泳馆：	7,759m²
训练馆：	4,017m²
体育场建筑：	1,069m²
看台：	2,630m²
建筑占地面积：	18,018m²
容积率：	0.27
绿化率：	34.8%
设计时间：	2005年

BASIC INFORMATION

Sports Center	
Total Base area:	95,300 m²
Total Building area:	26,250m²
Including	
Gymnasium:	10,775m²
Swimming hall:	7,759m²
Training hall:	4,017m²
Stadium:	1,069m²
Stands:	2,630m²
Site area:	18,018m²
Plot ratio:	0.27
Green ratio:	34.8%
Design time:	2005

(03)

Sports Centre

Sports Center is located in south part of the campus, north of the crossway of B1 highway and Luchao River, northeast of the teaching complex, with a total area of 24,000 square meters. The center contains a gymnasium, a swimming hall, a training hall and a ground track field.

As a whole, the gymnasium, swimming hall and ground track field stand in accordance with an arc facing the center of the campus and enclose an open oval square, which meets the needs of big performances and assemblies and embodies the character of Sports Center.

Architecturally, the three relatively independent building - the gymnasium, swimming hall and training hall centripetally connect each other with an angle and naturally forms a semi-open and enclosed space, which indicates the huge tension of sports center. The shape of the halls come from the idea 'origin from drips and converge at oceans'. The three halls emerge as three drips of different sizes in a converging state, by using the method of skin transformation. This indicates a philosophy: small drips go into immense oceans through countless convergence.

These two cases are two of the few projects that are not carried out in this book. But the architectural esthetics inside them is embracing grace that we have to remember. So we record it in this book in order to console ourselves.

(05)

03 体育中心的设计取意“古代海轮”，预示着海事人进取拼拼搏的大海精神。

The concept of Sports Centre is coming from old ships; it represents SMUer's sea spirit of promotion and struggle.

04 总布置表达出了“源之点滴，瀚于海洋”的设计总构思——“三滴正欲融合的水滴”。

The plane arrangement follows the idea that 'origins from drips and converges at oceans'. The three halls emerge as three drips of different sizes in a converging state.

05 立面造型反映出了“舰船”的形象符号，结合建筑的功能设计了“船窗”、“甲板”、“铆钉”等构件，丰富了建筑的造型。从卢潮落对岸看过来，体育中心像似一个即将起航的“巨舰”。

The outside of the sports centre remind people of ships. Incorporating its functions, We also designed ship windows boards rivets and etc on the facade so as to make it real. Seen from Luchaogang, the sports centre resembles a big vessel ready to sail.

06 夜暮下的“体育馆、游泳馆、训练馆”紧紧融合在中心交往核心周围。两个大馆“体育馆与游泳馆”用一种开放的大姿态面向体育中心主广场。

At dusk, the gymnasium, the swimming hall and the training hall is closely enclosing the center, which serves for exchange and communication. The two big halls--the gymnasium and the swimming hall are standing open to the central square.

07 08 09 工作模型，闪烁着灵感的火花。

Models. The inspiration is sparkling.

07

08

09

The East Campus, Shanghai University

上海大学东部校区

Shanghai, China, 2005.

新校区建成10年来，通过校区布局结构、专业结构调整和学科优化重组，跻身于全国先进高校行列。随着教育事业的快速发展，新建了新校区，但由于办学规模的扩大，急需重新规划拓展校园建设空间，以适应建设综合性研究型大学的需要。上海大学东部校区在宝山区大场镇，基地面积约33万m^2，与校本部仅一路之隔。正确处理东校区与校本部的整体关系，合理规划东区校园与周围道路交通、后期开发用地的关系，营造出符合现代化要求并具发展潜力的有特色的校园场所，是本设计中特别关注的问题。

东部校区与校本部隔街相望，两个校区用地相对独立，连接薄弱。原有校区已经形成了以图书馆为中心的格局。东区作为校区的延续和发展，以低调的姿态生长并依附于原有中心，并且与校本部联系方便又各得其所。由于校本部基地脉络以图书馆为中心展开，道路以及建筑的形态形成了与入口相联系的斜向轴线关系，独立于城市之外形成独特的肌理形态。而建筑是传承着记忆的片断，作为文化的载体，是一种特殊的“文化物种”，既要传承地方文脉，又要外溢时代精神，与原有校区和城市之间都要达到整合统一。因此，在东区的设计中着重梳理两校区之间、校区与城市之间的肌理关系，使东区既与校本部相映成整体，又能自如地契入城市的背景中成为其有机生长的群体，从而使两校区成为城市图底上一个跳跃的片断而不是肢解的个体，一个活跃元而不是异化的元素。

01 上海大学是一所快速发展的新兴大学，建筑也以简洁现代的风格表征其开放进取的校风。
Shanghai University is a new rapid-developing university. Its simple modern buildings indicate its school spirit of enterprise and openness.

Since its establishment 10 years ago, Shanghai University has become one of the best universities in China through campus layout, faculty adjustment and subject reallocation. A new campus has been built to cater for the rapid development of education. However, with the increasing scale of education, it is urgent to re-plan the construction of the campus area to meet the needs of building a comprehensive university. The East Campus of Shanghai University (330,000 square meters), is located in Da Chang Town, Bao Shan District, and it is one street away from the Main Campus. This design particularly focuses on how to handle the overall relationship between the Eastern Campus and the Main Campus, how to plan the relationship between the Eastern Campus and the nearby road system and undeveloped land, and how to create a modern, promising and unique campus.

The East Campus is only one road away from the Main Campus. The two campuses are relatively independent, with few connections. The Main Campus is centered on the library. The East Campus, as the extension of the Main Campus, relies on the original center and has easy access to the Main Campus. As the Main Campus is centered on the library, the nearby roads and buildings have formed an inclining axis with the entrance. Buildings bring out the segments of memory, serve as a carrier of culture, and are special 'cultural species'. It should not only carry on the local culture, but also exemplify the spirit of the time, so as to unify the new campus with the old one and the city. Therefore, the design of the East Campus should handle the relationship between the two campuses and the relationship between the new campus and the city so that the East and the Main Campus can be harmonized with each other, the East Campus can naturally become part of the city environment, and the two campuses can become a vivid part of the city panorama instead of separate individuals.

02 手绘鸟瞰效果图。
Airscape by sketch.

基本资料

地理位置：	上海市宝山区
用地面积：	331,279㎡
总建筑面积：	245,570㎡
建筑密度：	18.1%
容积率：	0.74
绿化率：	40.32%
设计时间：	2005 年

BASIC INFORMATION

Location:	Bao Shan District, Shanghai
Base area:	331,279 m^2
Building area:	245,570 m^2
Construction density:	18.1%
Plot ratio:	0.74
Green ratio:	40.32%
Design time:	2005

Fenyang Campus, Shanghai Conservatory of Music

上海音乐学院汾阳路校区改造规划

Shanghai, China, 2004.

上海音乐学院创建于1927年，是一所历史悠久，享誉海外的音乐学府，有“音乐家摇篮”的美誉。上海音乐学院汾阳路校区坐落于今上海市淮海中路，是市中心繁华地段一处幽雅别致的“净土”，是上海三大商业街之一——淮海路——高雅品位的象征。

上海音乐学院创校77年，人文荟萃，物宝风流，涌现出许多音乐名家，校内的专家楼原为上海犹太人俱乐部，音乐会所原为比利时领事馆，行政楼也是一座大方典雅的古典花园洋房，这些建筑都体现出上海音乐学院深厚的历史人文底蕴。

整个园区内，郁郁葱葱的绿化作为学院的一个特征元素而显得十分突出，树木蔓延浸透至校园的结构网络内，绿化了校区并成为校区面貌的主要特征。

学院内随历史不断增添的建筑物之间关系暧昧凌乱，组织结构松散。新上海音乐学院规划设计处于城市中心，是与整个社会、文化、经济活动紧密结合的校园，提倡的不是封闭，而是交织；不是独立，而是融合。上海音乐学院利用了其优越的地理位置，在优化传统功能分区的基础上，增加对外的交流与共享区域（市民音乐厅及乐器博物馆），注重校园空间与城市空间的互动关系，无论是功能还是空间结构都成为了城市的一个有机组成部分，并且对于社会有了充分的开放度。

设计对上海音乐学院就地改造，扩大建设用地，对现有校区重新规划，调整优化功能分区，拆除和改造部分老建筑，新建部分新建筑，实现其具有艺术性、国际性、时代性相融的一流精品音乐学院的目标，并发挥学院提升区域文化氛围的作用，形成与上海历史风貌保护区相适应的上海音乐文化中心。

01 改造设计后的校园生态中心区。
The centre ecological area after reconstrution.

02 改造设计前的校园生态中心区。
The centre ecological area before reconstrution.

基本资料		BASIC INFORMATION		
地理位置：	上海市徐家汇	Location:		Xuhui District, Shanghai
用地面积：	48,000m²	Base area:		48,000 m²
建筑面积：	61,897 m²	Building area:		61,897 m²
原有建筑面积：	54,712m²	Original building area:		54,712 m²
其中：保留建筑	31,797 m²	Including:	Remaining areas:	31,797 m²
拆除建筑	22,915 m²		Pulling down areas:	22,915 m²
地上新建建筑面积：	30,100 m²		New areas on the ground:	30,100 m²
地下新建建筑面积：	11,000 m²		New areas underground:	11,000 m²
容积率：	1.29	Plot ratio:		1.29
绿化率：	35.7%	Green ratio:		35.7%
设计时间:	2004年	Time:		2004

1、Green land
2、Entrance of Theater
3、Entrance of Garage
4、Entrance of Pedestrian
5、Entrance of Underground Museum
6、North Entrance of College
7、Sunk Courtyard
8、Plaza
9、Park Lot
10、Centric Plaza
11、South Entrance of College
12、Plaza

East Lake Hotel

Huai Hai Road

Fen Yan Road

Xing Fu Road

East China Normal University
Huaihai Road Campus

New Construction:
A1 Theater
A2、Activity Center
A3、Teaching Complex
A4、Appending Building
A5、Refectory
A6、Playground

Original Construction:
B1、Music Chamber
B2、High-rising
B3、Music Center
B4、Administration
B5、Experts' Complex
B6、Teaching Complex

Site plan / 总平面

03 高层音乐专业教学楼是学校20世纪90年代的建筑，设计着手对其外表皮采用古典三段式的手法进行改造，使其与音乐学院的特殊地位相融合，成为音乐学院在该地区的标志性建筑。
The tower of professional music classroom was built in 1990s.We tried to renew it by decorating its skin with classical three-segment pattern. Through our effort, we hoped it would accord with the particular status of SCM and become the symbolic building in that area.

04 与地域性格背道而驰的原高层教学楼。
The picture of original tower, didn't accord with the area character.

Shanghai Conservatory of Music was established in 1927, which is called 'the cradle of musicians' and has won high reputation in the world. Located in Middle Huaihai Road, Fenyang Campus of Shanghai Conservatory of Music has become a pure land in the city center and the symbol of elegance at Huaihai Road.

With a history of 75 years, Shanghai Conservatory of Music is possessed of abundant humanities, culture and treasure troveare. Thousands of outstanding musicians have welled up from the conservatory. On campus, Expert's building was the original Shanghai Jewish Club. Music Club was the original Belgian Consulate, while Administration building was a big elegant classical villa. All of these buildings embody the deep history humanities of Shanghai Conservatory of Music.

On campus, the blooming green land plays a very important role. The trees are spreading to soak through the structure network and becoming a main character of the campus.

With the continuous increase of new buildings, the school structure is considered ambiguous and disordered. The new planning of Shanghai Conservatory of Music is located at city center with a close relation with the society, culture and economic activities. We promote interweaving but not isolation. We advocate fusion but not independence. Taking good advantage of its superior geographical location, the conservatory is planned to increase the area for communication and share (civil music hall and musical instrument museum) and pays much attention to the exchanging relationship between school space and city space, on the basis of optimizing traditional function areas. Thus, the space structure and function structure become organic parts of the city and provide enough openness to the society.

In order to establish an international first-class conservatory with high reputation of art and time, to promote the cultural influence of the district and to build a music center allied with the history-protecting areas in Shanghai, we plan to enlarge the construction site, re-plan the school, adjust and optimize function areas, pull down and revitalize some old buildings and put up some new buildings.

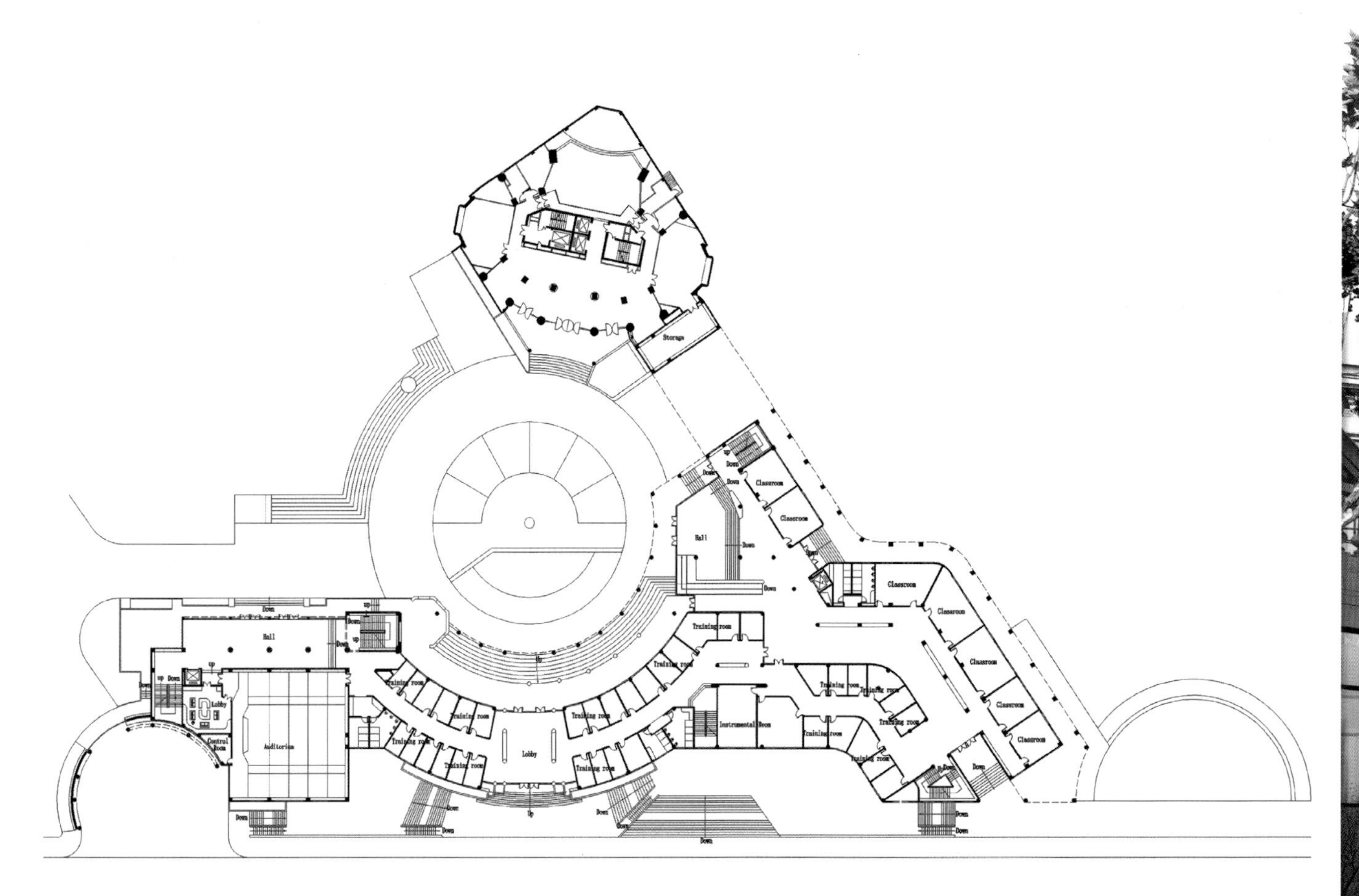

Fisrt floor / 教学楼一层平面

05 高层的教学楼连同新建的多层教学楼共同围合出音乐学院的建筑群圆形形态中心，面向校区内部相对打开，促进交流；面向相对更为对外的音乐厅共享交流中心用一个二层高的拱门连通。

The tower cooperating the newly-built, multi-storied teaching building encompasses a central road area of the complex. It is relatively open to the inside in order to promote communication. The music hall and exchange centre which is relatively open to the outside, a two-story arch door.

05

06 音乐厅与地下东方乐器博物馆的入口庭院空间，书写了新音乐学院高贵典雅的建筑特色，在闹市中别有一番“小资风情”。

07 电脑虚拟下的灯火辉煌的音乐厅，显示出音乐学院所处的上海淮海路高尚商业街区的地段魅力，上海音乐学院音乐厅正用她凝练的建筑形象和音乐会的特殊气氛吸引着人们纷至沓来。

The computer-made picture of SCM at night , indicates the charming of Huaihai Road, a gracious commercial street where SCM is located . The music hall of SCM attracts people by its elegant image and special atmosphere.

08 设计之初，对经历70年代的老校区保护树木做了一个完整的GPS定位，并且做出了新规划校区的保护树木移栽策略图。

At the beginning of the design, we made an accurate GPS position for all the 70-year maintaining plants on old campus. Then, we drew up a plan for the transplanting plants on new campus.

07

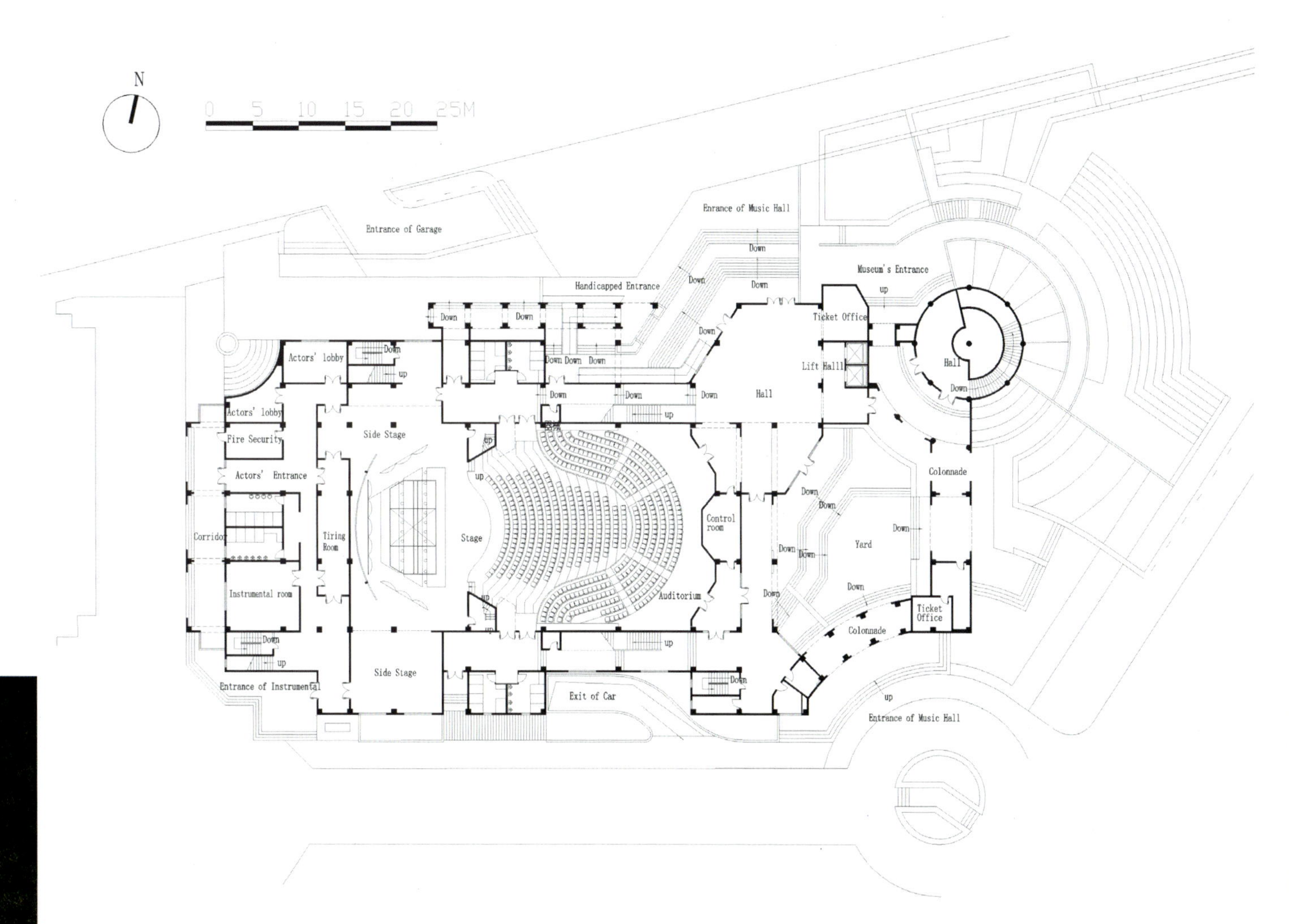

First floor / 音乐厅一层平面

原有校区保护树木定位图

Position image of the maintaining plants on old campus

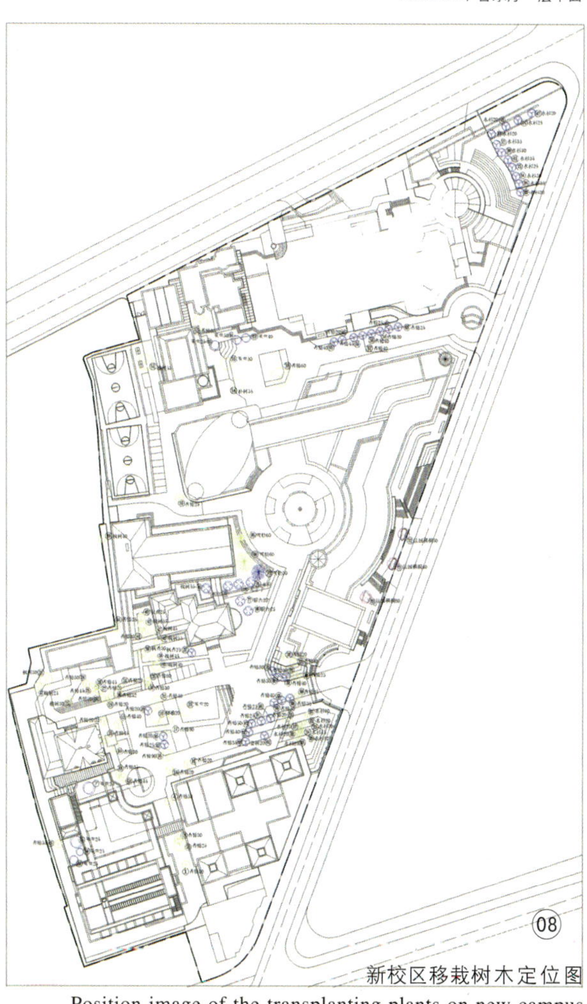

新校区移栽树木定位图

Position image of the transplanting plants on new campus

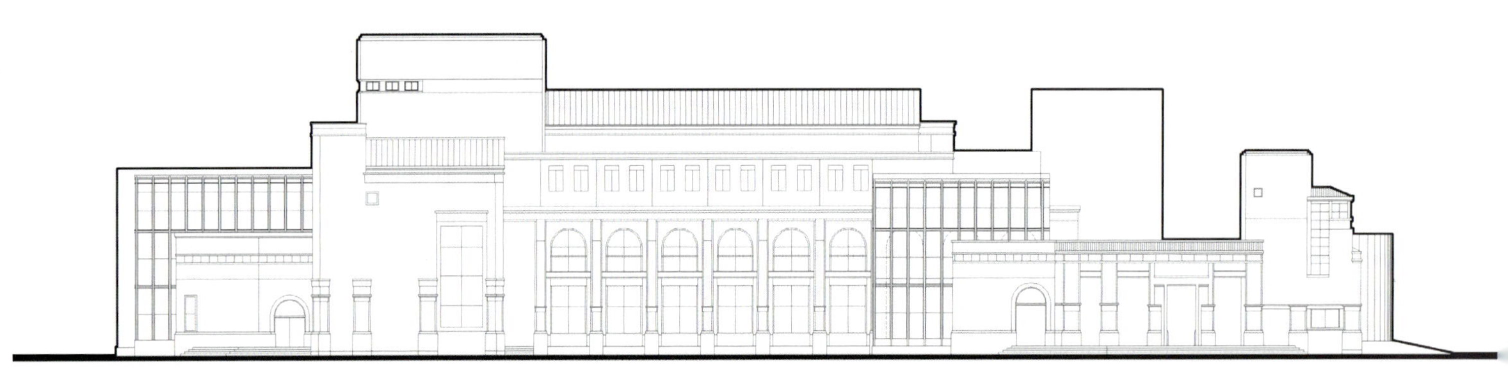

Facade / 音乐厅沿街立面

09 改造后保留了建筑间古树的音乐会所与活动中心。

The reconstructed Music club and activity centre, where the old trees are preserved.

10 与校内历史保护建筑原上海比利时领事馆和谐共生的新建音乐厅，正如上海这座城市一样，焕发着别样的“花样年华”。

The newly-built music hall incorporating preserved original Belgian Consulate is blooming like the City of Shanghai.

11 改造前的音乐会所。

The music club before reconstruction.

12 改造前后电化教学楼。

The informative teaching building before and after reconstruction.

13 一颗璀璨的音乐明珠在上海的中心散发着夺目的光辉。
A bright music pearl is twinkling in the centre of Shanghai.

接近古典理想

——上海音乐学院设计随笔

学校是这样一类建筑物，他们不仅是出于功利或需要，也不仅仅是为了提供栖息之所，它是为了某种崇高的目标而存在，需要某种风格的内涵且尊敬之，改造前的上海音乐学院显然缺乏这种直指人心的力量。

一直以来，学生们通过视觉和感官的信息不断地获得对学校性格和文化的了解，而学校的景观便是信息的源泉，也就成为学生产生集体认同和记忆的基础。上音应具有其他学校无法代替的个性，如何找到它，这是一个艰难的过程。

上音的建筑应不仅仅感受到物质的“美”，还有校园的历史、文化、氛围等种种载体，折射出的是校园历经百年逐渐完善的文化内涵和历史印记。现代主义理论是建立在对建筑物质文化的诠释基础之上的，其价值判断具有鲜明的条理化倾向；历史主义则以建筑的精神文化内涵为其理论核心，价值判断展示出浓烈的人文关怀与人本思想。我们逆着西方美术史的时针回溯，从文艺复兴一直到古代希腊和罗马的古典艺术，那种“高贵的单纯和静穆的伟大”的古典精神，深深吸引着我们，于是，历史折衷主义的法则被明确为上音校园的价值取向。

但在奉现代主义为圭臬的建筑设计界，认真于历史新古典的风格，颇有点冒天下之大不韪的感觉，然而，设计本不该有什么禁区，至少我们是如此认为。

上海音乐学院的规划特征主要关注的是建筑形体的清晰，视觉的领会性和空间的场所感，强调张弛有致的节奏和韵律，空间规划则主要注重功能空间的有机组织和开放空间的连续展开。场所精神是设计师的时髦语汇，但很少有人可以说清楚究竟是什么意思，我们理解为主导校园气氛的某种空间感觉，而非场所本身，他向人们表达一种整体的意象，具有舒适、安全、依赖等种种含义。这对于一个城市中心的袖珍型校园，是极为关键的。

当原有基地中的待拆建筑拆除后，呈现于我们面前的是一条保留建筑复杂曲折却又有机参差的界面，为尊重这条界面的形态，新规划的内侧界面蜿蜒曲折，时而与原有界面吻合，时而又与原有界面退让，很好地呼应了原有校园的空间肌理。所以，上音校园与城市，道路之间还有许多你中有我，我中有你的渗透空间，这种空间上的暧昧，消弥了学校与城市之间的界线，以开放的姿态欢迎感悟音乐，享受音乐，追求音乐的人们。

步行由淮海路向南，走过音乐会所与音乐厅之间尺度宜人的狭长台地；车行则由汾阳路北入口向西驶过漫漫林荫道，便可见到在活动中心、音乐会所、实验剧场（博物馆）与新建教学楼环绕下的开阔广场。广场中心的一棵参天巨木是上海音乐学院原状保存的老香樟树，她位于淮海路视觉走廊、汾阳路北入口视觉轴线、新教学楼拱门框景轴线的端点上，是上海音乐学院规划变迁的原点，见证着历史沧桑的巨变。

广场将为活动中心中的上音师生，实验剧场内的座上嘉宾，音乐会所中的同道乐迷，流连于博物馆内的参观游客，以及周围社区的居民们提供日常休闲的集散地。她是属于上海音乐学院的校内空间，却又仿佛是音乐学院的校前广场，因为走过新教学楼的大拱门，才真正进入上海音乐学院的教学行政区；这个共享空间也似淮海路上以文化艺术设施为主打品牌的空间节点，连接剧场、博物馆、会所与艺术院校，所以她更属于这条街道，这座城市。

音乐会所与剧场之间形成新天地般怡人古雅的里弄尺度，浓浓的树荫下，人们休憩闲坐。同时形成视觉走廊，将淮海路上的人们视觉焦点引入上海音乐教学楼的大拱门前，使得即便在繁华的、热闹的淮海路上，人们也能一窥音乐的净土。

上海音乐学院汾阳路北入口，同时也是博物馆与剧院的车行入口，与南入口同为圆形平面，但管理更为开放自由，将人们由林荫大道引至拱门前，刻意弱化自己的入口形态，强化拱门的入口形式，仿佛经由一系列空间序列终于来到了校前广场，见到了拱形校园大门。

走过宏伟的拱形门廊，进入圆形广场，映入眼帘的便是气魄惊人的教学楼群组。我们可以通过复调法或和声获得深度，赋予音乐以翫峁箹，建筑也如是：当我们把高层教学楼与新教学楼连接起来，她们演绎出完美的和声，如同巴赫的《平均律钢琴曲》，以节奏的强烈与形式的清晰，比例的壮观，体现出凝练而富有个性的主题和富于想象建筑形象，每一个细节都体现出结构上的完美技术。同时他们形成了一个宏大完整的圆形广场，它发挥着主导校园气氛的重要作用，向人们提供一种整体的意象。圆形的形态中心处于两栋教学楼的中心，成为日常学习和生活的延伸，社会功利的一面在此庄重的人性色彩中获得稀释，她以一种向四周通透开放的姿态，欢迎学生参与到这个环境中来。

由汾阳路南入口（上海音乐学院原有入口）向西进入学院，展现于眼前的是一大片保存完整的草坪与郁郁葱葱的树林，树林后隐约可见专家楼绰约的风姿。大片的绿地、简单的设计，绿地中点缀以音乐名家的雕塑，使这片广大、纯粹、安静的绿地成为上海音乐学子们感怀沉思的场所。

漫步在上音的校园，暗红色的砖墙和浅色石料在斑驳的树影之中默默无言，让人安静内敛，感怀深思。古典元素已成为音乐人精神结构中的一种历史情结，一种埋藏在心灵深处的原形图式，只要遇到合适的环境和温煦的阳光，这颗文化与艺术的种子就会生根发芽。这也正是上音的新建与改建设计的原创动力和内在母题。

当建筑以三段式的比例，精巧的细部、拱券、回廊在历史与现实之间建立了一种文脉上的联系，并产生了强烈的修辞效果时，她已不是超越历史时空的工业理性，而是后工业时代的一种有厚度的形式美，一种度身定做的契合感，一种历史感，一种文化纵深感，一种如此贴切的形式感和发自人内心的归属感，你将看到一种沉甸甸的东西，一种文化意蕴，一种精神的归属。

——后记

2004年，夏天刚刚过去，秋热的余燥未消，但是公司的空气中激荡着某种兴奋——我们接到了一个特别的项目：上海音乐学院汾阳路校区改扩建规划建筑设计。回望公司的发展史，正因为这个项目，一批新人被承认为设计骨干，当年那个精干的设计小组，里面的每一个成员如今都已成为公司独当一面的带头人，然而，这个项目可能是公司唯一一个势在必得却在最后关头失手的投标。上海音乐学院的失败也成为一种情结，萦绕在每个参与者的心头，成为难以宣之于口的隐痛，痛苦于它的美藏在深闺无人得知，痛苦于自己呕心沥血的作品被束之高阁。于是，有了上面一文以纪录曾经的设计历程。在设计上海音乐学院的过程中，被激情点燃和照亮的，是一群年轻建筑师心中蛰伏久已的对永恒和崇高的深深敬意，这支持我们一直前行。

项目主创人员注引　Architect Index

后 记

历史是一位与所有人签订了生死契约的不死老者，他的顽强、狡诈和吊诡超过一切凡人。他以谁为代言，在什么时候发声，以何种形态返回尘寰，不可蠡测。

罗凯先生，湖南长沙人，行将度过其第三个本命年。他以宽广黑黝的额头、深邃坚毅的双眼、诗人的灵心善感，哲人的敏察善思，为建筑设计的价值判断思虑万千。他犹如一名预告坍塌和毁灭的隐修士，唱诗班中永不现身的梦幻幽灵，送葬行列中远远驻足沉思的局外人，眨着冷眼颤着热心，用炼丹术般的修辞和锤炼了千遍的建筑修辞，为中国高校建筑设计壮行。

在人海里载浮载沉，才知道日子原是不偏不倚、亦步亦趋地平分给我们每一个人。生活规则里刻琢出来的我们好像有着同样的面纱和盔甲，就像一条蚕轻轻爬在生活的网格里给自己作茧。市井的纷繁在人与人之间隔出一道墙，再难感觉那种真纯得可以见底的心灵。

我与罗凯先生相交经年，在我们有限的接近中，笼罩在他身上太多的忧郁和孤寂，一种与生俱来的悲情和不忍之心使我惊讶、揪心。在他的身上，总有老式的幻灭，簇新的愤怒，一轮又一轮的精神炼狱，几乎没有任何排遣、消闲。他那天生的忧世伤时比寻常的幽愤更苦涩，比个人的不幸更绝望，比通常的悲观厌世更尖锐，他所唤起和拒绝的有时甚至超过了虚无，找不到任何与之相对的物事。他的敏感、同情、正义和英勇，他的才华、血性、灵气、悟性，直接预示着个体生命精神破晓的壮丽。

我可以想像他这几年的光阴是如何度过，孤灯清卷，夜不能寐，在奔忙的岁月中，寻找在他心中奔泻的创作梦想。他的建筑作品饱蘸了对历史和文化眷恋的激情，凝结着对文化教育建筑设计的感悟和思辩，充满洒脱与不羁，转述和表达着对建筑文化和艺术史的顶礼膜拜；他的态度是笃定的，眼神里布满宗教徒的虔诚 —— 一种对建筑文化和历史负责的虔诚；经建筑实践造就了他硬朗而霸道的设计思维，在喧嚣嘈杂而更需要一种清晰的方向感的时候，他以显扬的设计笔触苦心孤诣地梳理着建筑历史观和人文情怀，传承在颠覆中隽永依旧。睿智和执着让他的探索和奋斗固化出一层光晕，使他在教育建筑设计方面占有了一席之地，也更坚定了他的选择的理智——甚至方向。"

创新真的存在吗？也许正因为创新可以是许多东西，也许确切的创新根本从未存在过，他的作品诚如他在一偏文章中所说的"你可以赞美她的深厚，也可以批评她古典，你可以抨击她的孤傲，但你无法否认她所具有的深厚的文化沉淀和动人心魄的精神依恋"。所以他并不期望从创作中获得什么所谓的思想总结，建筑创作仅仅是他的一种生活方式，建筑创作已经成为他生活的一部分。正如罗凯先生自己所说："我思想，所以我存在"。

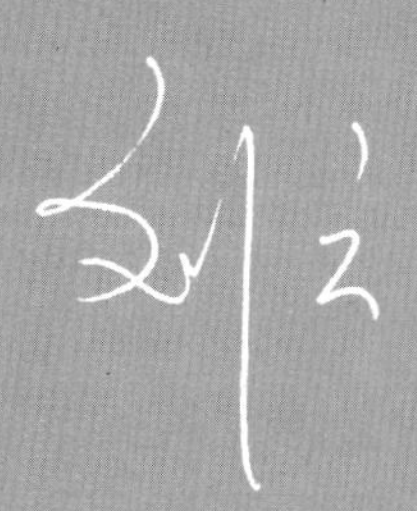

上海华东建设发展设计有限公司设计总监

2006-11-3

Postscript

History is an immortal that signed a death and life contract with people. His brawniness, slyness, and oddity exceed all mortals. Who does he advocate for and when? In what form does he return to earth? All these questions are unanswerable.

Mr. Larkin, from Hunan Changsha, is living through his 36th year. He thought hard about the value of architectural design with his broad, swarthy forehead, profound eyes, sensitive heart of a poet, and reasons and observations of a philosopher. He is like a friar that has been warned of the destruction of his church; a phantom that never showed up for the choir; an outcast who is absorbed in his own thoughts at a funeral. His extensive architecture rhetoric helped shape the higher education of architecture in China.

In this life of ups and downs, days still go on no matter what happens. Throughout our life experience, we all seem to be wearing a mask, like a caterpillar wrapping itself in a cocoon, protecting ourselves in this complicated, urbanized world. Invisible walls form between people, so that it becomes difficult to once again feel the pureness of a soul.

I have known Mr. Larkin for several of years. In our limited contact, his solitude and sorrow took me by surprise, and filled my heart with grief. His life lacked leisure and entertainment. He was constantly going through cycles of psychological torment of disappointment and rage. His depression was even bitterer than our frustrations, more hopeless than any other misery, more extreme than any pessimist or cynic. What he sought might be considered to be nothing, and he was unable to find anything relevant. His sensitivity, compassion, righteousness, bravery, talent, courage, comprehension, and life signified the magnificence of the first ray of sunlight in life.

I can imagine how he spent the past few years. In his hectic life, he searched for his overwhelming dream of better creations through sleepless nights and endless work hours. His architectural works fulfilled his passion for history and culture; coagulated his thoughts and analysis of cultural education of architecture; expressed his devotion to architectural culture and art history. He is a determined person. His eyes reflect his devotion towards architectural culture and history like a religious person towards his God. Using an obvious design style, he struggled to hackle architectural history and intellectual thinking, while still keeping his style through the subversion of legacy. His wisdom and persistence transformed his hard work into glorious accomplishments and earned him an outstanding achievement in the education of architectural design, and also confirmed the wit of his choice and even direction.

Does innovation really exist? Maybe because innovation can be many things, that absolute innovation has never really existed. His works are like what he said once in an article, 'You can praise her profundity, and criticize her classicality. You can deprecate her pride, but you can't deny the profound cultural sediment she possesses and the soul-moving spiritual devotion.' Thus, he never expected to achieve the so-called 'conclusion' from each of his works. Architectural design has become a part of his life. Just like Mr. Larkin said, 'I think, then I exist.'

Yun Liu
Nov 3rd, 2006

图书在版编目(CIP)数据

历史·文化·传承——华谏国际文化教育建筑设计作品专辑/罗凯著.—北京：中国建筑工业出版社，2007
ISBN 978-7-112-02087-4

Ⅰ.历... Ⅱ.罗... Ⅲ.高等学校—建筑设计
Ⅳ.TU244.3

中国版本图书馆CIP数据核字(2006)第162489号

华谏国际建筑设计集团在2001—2006年的短短几年间，设计完成了大量的文化教育建筑与规划设计作品，提出了“新古典、现代感”的总构思，在建筑细部上认真仔细的推敲，以这样的方式彰显“历史·文化·传承”，得到了社会和舆论的认同。本书介绍了华谏国际建筑设计集团以高校规划与文化教育建筑设计为代表的一系列作品，以及作者所思、所想、所感的多篇文章，展现了新一代建筑师探索特定背景下的建筑设计思想的过程，对广大建筑师如何赢得市场和业主，收获丰富的成果，有良好的交流学习的作用。

责任编辑： 陈 桦

策　　划： 张约翰 杨申茂

装帧设计： 胡 萍 邱吉尔

编委成员： 张约翰 杨申茂 谢红妹 皮岸鸿 刘 云 任 凭 李 鑫 向卓睿
金万贤 朱 灵 凌颖松 杨 琳 李 盈 杜 凯 罗翌弘

摄　　影： 高雪雪

历史·文化·传承——华谏国际文化教育建筑设计作品专辑
罗凯 著
*
中国建筑工业出版社 出版、发行（北京西郊百万庄）
新华书店经销
北京中科印刷有限公司印刷
*
开本:880×1230毫米 1/16 印张:$17^{3}/4$ 字数：562千字
2007年2月第一版 2007年2月第一次印刷
印数：1—2000册 定价：198.00元
ISBN 978-7-112-02087-4
(14451)

本社网址：http://www.cabp.com.cn
网上书店：http://china-building.com.cn